revise GCSE

AQA

Author - Bob Hartman

Mathematics

Contents

 **Number**

 **Algebra**

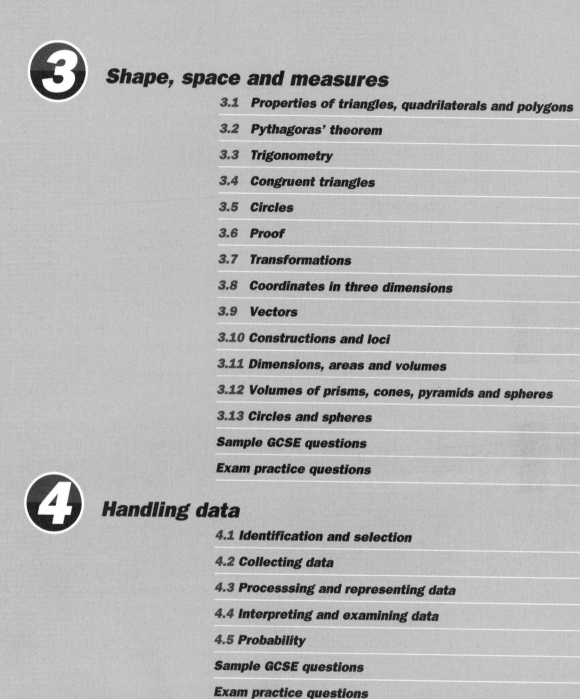

3 Shape, space and measures

4 Handling data

Your GCSE course

AQA offers two mathematics GCSE specifications, Mathematics A and Mathematics B. Each has two entry tiers, Foundation and Higher.

The tiers have a different range of possible grades, but with some questions in both tiers.

- **Foundation Tier**: Grades C, D, E, F, G or U with 50% of the marks for grade F or G material, and between 25% and 30% for grade C or D material.

- **Higher Tier**: Grades A*, A, B, C , D or U with 50% of the marks for grade C or D material, and between 25% and 30% for the two highest grades or B material.

To give you a *very rough* idea of question difficulty and grades here are a few examples of questions and their *possible* grades.

Grade G: Use a word formula, such as Cost of hire in £s = days hired × £20 + £10

Grade F: When $a = 2$, find the value of $s = 3a$

Grade E: Solve the equation, $3x - 4 = 5$

Grade D: Solve the equation, $3(2x + 1) = 21$

Grade C: Use trial and improvement to solve the equation, $x^3 + 2x = 20$, to 1 d.p.

Grade B: Solve the simultaneous equations, $4x - y = 10$ and $3x + 2y = 13$

Grade A: Solve the quadratic equation, $x^2 - 3x - 4 = 0$, using the quadratic formula.

Grade A*:   Solve the simultaneous equations, $y = 2x - 3$ and $x^2 + y^2 = 6$.

The **AQA Linear specification, Mathematics A (4301),** has the same coursework requirements as Mathematics B, which is two tasks together worth 20% of the total marks, one a mathematical investigation, the other testing data handling skills. There are two written papers, lasting $1\frac{1}{2}$ hours for Foundation Tier and 2 hours for Higher Tier. Paper 1 is non-calculator and Paper 2 calculator. Each is worth 40% of the total marks. All questions are compulsory and involve questions on Number, Algebra, Shape, space and measures and some Handling data.

The **AQA Modular specification, Mathematics B (4302),** consists of five modules. The content of the modules is different for each tier, but there will be some questions that are in both tiers. The way the modules are arranged does not differ between tiers of entry.

Module 1 (M1) tests the Handling data part of the specification. It accounts for 11% of the total marks. It is a written paper, consisting of two sections. Section A is non-calculator and Section B calculator. Each section takes 25 minutes and is marked out of 20. All questions are compulsory.

Module 2 (M2) is the Handling data coursework task. It accounts for 10% of the total marks.

Module 3 (M3) will mainly test the Number and Algebra parts of the specification. It accounts for 19% of the total marks. There are two sections. Section A is non-calculator and Section B calculator. Each section takes 40 minutes and is marked out of 32. All questions are compulsory.

Module 4 (M4) is the second coursework task. It accounts for 10% of the total marks.

Module 5 (M5 – also known as the terminal module) consists of two papers, each taking 1 hour 15 minutes and together accounting for 50% of the total marks. Paper 1 is non-calculator and Paper 2 calculator, each are marked out of 70. Both papers assess mainly Algebra and the whole of Shape, space and measures. All questions are compulsory.

Preparing for the examination

Planning your study

The final three months before taking your GCSE examination are very important in achieving your best grade. However, your success can be assisted by adopting an organised approach throughout the course.

- After completing a topic in school or college go through the topic again in *Revise GCSE Maths Study Guide*. Copy out the main points, results and formulae, etc. on a sheet of paper or use a highlighter pen to emphasise them.
- A few days later try to write out these key points again from memory. Check any differences between what you originally wrote and what you wrote later.
- If you have written your notes on a piece of paper keep this for revision later.
- Try some questions in the book and check your answers.
- Decide whether you have fully mastered the topic and write down any weaknesses you think you have.

Preparing a revision programme

At least three months before the final examination look at the list of topics in your awarding body's specification. Go through and identify which topics you feel you need to concentrate on. It is a temptation at this time to spend valuable revision time on the things you already know and can do. It makes you feel good but does not move you forward.

When you feel you have mastered all the topics spend time trying past questions. Each time check your answers with the answers given. In the final couple of weeks go back to your summary sheets (or your highlighting in the book).

How this book will help you

Revise GCSE Mathematics Guide will help you because:

- It contains the **essential content** for your GCSE course without the extra material that will not be examined.
- It contains **Progress Checks** and **GCSE questions** to help you confirm your understanding.
- It gives **sample GCSE questions** with **model answers** and advice from examiners on how to improve.
- The questions in this book have been written by experienced examiners.
- **Marginal comments** and highlighted **key points** will draw your attention to important things you might otherwise miss.

Five ways to improve your grade

1 Read the instructions carefully

Some of the instructions could be:

- **Answer *ALL* the questions.**
  This means answer as many as you can. Only by getting all of them right will you obtain full marks.

- **Write your answers in the spaces provided on the question paper.**
  You should not use any other paper.

- In the exam you should **check** that you have been given the **correct paper**, that you know **how many questions** you have to answer on that paper and **how long** you have to do it. Try to spread your time equally between the questions. If you do this it will avoid the desire to rush the paper or spend too much time on some questions and not finish the paper.

- Many mathematics papers start with fairly straightforward questions which may be shorter than those that follow. If this is the case **work through them in order** to build up your confidence. Do not overlook any part of a question and **double check** that you have seen everything on each paper, look especially at the back page in case there is a question there!

- **Take time** to read through all the questions carefully and then start with the question(s) that you think you can do best.

- When there are about 15 minutes remaining in the examination then quickly check if you are running out of time. If you think that you will run out of time then try to score as many marks as possible by concentrating on the easier parts, the first parts, of any questions that you have not yet attempted.

2 Read the question carefully

- Make sure you **understand what the question is asking**. Some questions are **structured** and some are unstructured – called **'multi-step' questions** – and for these you will have to decide how to tackle the question and it would be worthwhile spending a few seconds thinking the question through.

- Make sure you understand key words. The following glossary may help you in answering questions:

  Write down, state – no explanation is needed for an answer

  Calculate, find, show, solve – include enough working to make your method clear

  Draw – plot accurately using the graph paper provided and selecting a suitable scale if one is not given. Such an instruction is usually followed by asking you to read one or more values from your graph.

- The number of marks is given in brackets [ ] at the end of each question or part question. This gives some indication of **how many steps** will be required to answer the question and therefore of what **proportion of your time**, you should spend on each part of the question.

3 Show your working and check your answers

- **State units** if required and give your final answer to an **appropriate degree of accuracy**.
- Write down the figures on your calculator and then **make a suitable rounding**. Don't round the numbers during the calculation. This will often result in an incorrect answer.
- Don't forget to **check your answers**, especially to see that they are reasonable. The mean height of a group of men will not be 187 metres!
- **Lay out** your working **carefully** and **concisely**. Write down the calculations you are going to make. You usually get marks for showing a correct method. (If you are untidy and disorganised, you might misread some of your own work and/or lose marks because the examiner cannot read your work or follow your method.)
- **Remember** that marks are given for the following:
    - using an **appropriate method** to answer a question
    - for **facts found** as you work through a question
    - for the **final answer**.
- **Remember** that if all that is written down is an answer and that answer is wrong you gain no marks. Once you have finished the paper, if you have any time left **check the work** you have done. The best way to do this is to work through the questions again.

4 What examiners look for

The examiners look for the following:

- Work which is **legible**, **clearly set out** and **easy to follow** and understand. Use a pen, not pencil, except in drawings, and use the appropriate equipment.
- That drawings and graphs are **neat**, and **graphs are labelled**.
- That you always **indicate** how you **obtain your answers**.
- **The right answer!**

5 Practice makes perfect

- **Practise** answering questions that ask for an explanation. Your answers should be concise and use mathematical terms where appropriate.
- **Practise** answering questions with more than one step to the answer, e.g. finding the radius of a sphere with the same volume as a given cone.
- **Practise** all aspects of manipulative algebra, solving equations, rearranging formulas, expanding brackets, factorising, etc.
- Make sure you can use your **calculator efficiently**.
- Look at some old mark schemes to give yourself a feel for how examiners award marks – some of these are available on the AQA website, as well as examiners' reports.

Coursework

You will have to complete two coursework tasks, one on Using and Applying Mathematics, and one on Handling Data. Each task will contribute 10% towards your final grade, leaving 80% of your marks to be earned from your performance on the written examination papers. Each task will be assessed using the appropriate criteria, with marks being awarded for performance in 'assessment strands', which are outlined in the table below.

There are two different methods for assessing the two coursework tasks that are completed. These are:

- Your school can select the tasks to be completed, mark all of them and send a sample of marked scripts to the awarding body to be moderated.
- The awarding body will set the tasks and mark the work.

Task type	Details	Weighting	Assessment strands	Max marks
Using and Applying Mathematics	One task, usually, based on number and/or algebra or based on shape and space.	10%	1 Making and monitoring decisions	8
			2 Communicating mathematically	8
			3 Developing the skills of mathematical reasoning	8
Handling Data	One task (which should be based on a statistics activity rather than on a probability based activity)	10%	1 Specify and plan	8
			2 Collect, process and represent	8
			3 Interpret and discuss	8

Thus the **maximum mark** for each task is **24**. The marks from each task that you gain will be added together to give a total coursework mark.

For example your marks could be:

	strand 1	strand 2	strand 3	Total
Using and Applying Mathematics	6	6	5	17
Handling Data	6	5	4	15
Final total				32

The detailed criteria for Using and Applying Mathematics and for Handling Data, which are common to all awarding bodies, are written by the QCA and printed in the specification that you will be using. You should be able to obtain a copy of these detailed criteria from your teacher or the awarding body's website.

There are some well-defined strategies that you might wish to adopt.

Using and Applying Mathematics Strand 1: Making and monitoring decisions

4 marks	Gathering (in a systematic manner) enough results that are correct and enable you to write a generalisation about the given problem.
5 marks	Change one variable and undertake sufficient new work so you could make a further generalisation.
6 marks	Show a range of techniques to extend and develop the task further. For example if you had only been using simple linear equations such as $y = 3x - 2$ up to this point you could try to use a graphical approach or simultaneous or quadratic equations to support this extended work. (This would link in with the requirement for 6 marks in the Communication strand where the consistent use of symbolism, i.e. algebra, is required.)
7 marks	Attempt to coordinate three features in the work, perhaps by moving into 3 dimensions.

Using and Applying Mathematics Strand 2: Communicating mathematically

4 marks	Present work in an orderly manner using two different methods, for example tables and diagrams, linking them together with a commentary.
5–6 marks	Show an increased use of algebra.
7–8 marks	Show a sophisticated use of algebraic techniques.

Using and Applying Mathematics Strand 3: Mathematical reasoning

3 marks	Show a progression from 'making general statements', i.e. a valid generalisation, derived from at least three of your results.
4 marks	Test your findings, formula or relationship by checking a further case (do not use the values you already used in deriving the formula or results).
5 marks	Give a valid explanation as to why your generalisation works, referring to the shape of a grid, or size and structure of a shape.
6–8 marks	The progression continues up to 8 marks where a mathematically rigorous justification is expected.

Handling Data Strand 1: Specify and plan

5–6 marks	Show clear aims and state a plan designed to meet these aims. The data used should be appropriate and the reason for any sampling should be explained.
7–8 marks	Demonstrate valid reasons for what you have done and explain any limitations, for example bias, that might arise.

Handling Data Strand 2: Collect, process and represent	
5–6 marks	Show correct use of appropriate calculations using relevant data.
7–8 marks	Demonstrate evidence of higher level techniques applied accurately.

Handling Data Strand 3: Interpret and discuss	
5–6 marks	Use summary statistics to make comparisons between sets of data and clearly relating your findings back to the original problem and evaluating the success, or otherwise, of your strategy.
7–8 marks	Explain how you avoided bias and demonstrate the use, for example, of a pre-test or a pilot questionnaire.

Tips

There are various techniques for you to use which should help you to improve your marks on both types of coursework.

One technique is to use a three part approach, and for each part think through some or all of the following questions, answering them in your head or on rough paper (but don't include them in your submission):

Part	Using and applying mathematics task	Handling data task
1 Starting a task	What does the task tell me?	Do I have a clear plan and clear aims?
	What does it ask me?	What questions do I have to answer? Is a hypothesis given or can I think of one?
	What can I do to get started? For example find a simple starting point	What data do I think I will need? When, where and how will I get it – doing a survey, using a questionnaire, doing an experiment?
2 Working on the task	What connections are possible?	Am I using the right technique?
	Is my method clear?	Is my method clear?
	Is there a result to help me?	Is the data relevant and appropriate?
	Is there a pattern in the results I have found?	Are some patterns, trends and/or conclusions beginning to emerge?
	Can the problem be changed?	Can I think of different questions to ask?
3 The review, conclusion and extension	Is the solution acceptable?	Have I used the most appropriate presentation?
	Are my results presented as clearly as possible with explanations and reasons?	Are my results presented as clearly as possible with explanations and reasons?
	Can it be extended?	Can I think of further questions to ask – perhaps as a result of my findings so far?
	What conclusions can be made?	Have I answered the starting questions or hypothesis?

Number

Overview

Topic	Section	Studied in class	Revised	Practice questions
1.1 Integers	Factors, multiples and primes	✓	25.04.07	
	Negative numbers	✓		
1.2 Powers and roots	Square roots and cube roots	✓		
	Laws of indices	✓		
	Inverse operations	✓		
	Standard index form	✓		
1.3 Fractions	Equivalent fractions	✓		
	Addition and subtraction of fractions	✓		
	Multiplication of fractions	✓		
	Division of fractions	✓		
	Mixed fractions			
1.4 Decimals	Decimals and fractions			
1.5 Percentages	Percentages and fractions			
	Percentages and decimals			
	One quantity as a percentage of another			
	Finding a percentage of a quantity			
	Finding a percentage increase or decrease			
	Reversed percentages			
1.6 Ratio	Simplifying ratios			
	Dividing in a ratio			
1.7 Mental methods	Number facts			
	Rounding numbers			
	Making estimates			
1.8 Written methods	Pencil and paper			
	Proportional change			
	Irrational numbers			
	Calculating with surds			
1.9 Calculator methods	Understanding your calculator's keys and display	✓		
	Exponential growth and decay			
	Upper and lower bounds			
1.10 Solving numerical problems	Strategies			
	Checking			

1.1 Integers

After studying this section, you will be able to:
- *find common factors and multiples*
- *recognise and use prime numbers*
- *use and understand negative integers*

Factors, multiples and primes

An **integer** is any whole number, whether positive, negative or zero: –4, 0 and 13 are integers.

> A common error is to confuse factor and multiple.

Factors

5 is a **factor** of 30, because 30 can be written as 5 × 'an integer' (30 = 5 × 6). 5 is **not a factor** of 33 because 33 cannot be written as 5 × 'an integer'. The factors of 20 are 2, 4, 5 and 10 and also 1 and 20 itself.

> **KEY POINT**
>
> A **common factor** is a number that is a factor of each one of a given set of numbers: 11 is a common factor of 22, 110 and 132. The highest common factor (HCF) is the largest number which will divide exactly into two or more other numbers: the HCF of 18, 21 and 24 is 3, the HCF of 12, 24 and 30 is 6.

Multiples

A **multiple** is a number that is a product of a given whole number and another whole number: 15, 40, 110 and 2005 are all multiples of 5; 4, –12, –60 and 300 are all multiples of 2.

> **KEY POINT**
>
> The lowest common multiple (LCM) is the smallest number that is a multiple of each one of a given set of numbers: for the numbers 2, 3 and 4, 24 is a common multiple, but 12 is the LCM.

Prime numbers

> **KEY POINT**
>
> A **prime number** is a positive integer which has exactly two factors – itself and 1: 1 is not a prime number; 2, 3, 97 are prime numbers. All positive integers can be written as a product of prime factors: 52 = 2 × 2 × 13.

Examples

(a) Write 28 and 60 as products of prime factors.

Hence find the LCM and HCF of 28 and 60.

2	28
2	14
7	7
	1

2	60
2	30
3	15
5	5
	1

$$28 = 2 \times 2 \times 7$$

$$60 = 2 \times 2 \times 3 \times 5$$

The LCM must include $2 \times 2 \times 7$ (= 28) **and** $2 \times 2 \times 3 \times 5$ (60) (by definition 28×60 must be a multiple of both 28 and 60 – but not necessarily the lowest) but 2×2 occurs in both, so the

$$\text{LCM} = 2 \times 2 \times 3 \times 5 \times 7 = 420$$

The only common factor is 2×2, so the HCF is 4.

(b) Use the fact that

$$84 = 2 \times 2 \times 3 \times 7 \text{ and } 90 = 2 \times 3 \times 3 \times 5$$

to find the LCM and HCF of 84 and 180.

A common multiple (but not necessarily the lowest) is

$$84 \times 90 = (2 \times 2 \times 3 \times 7) \times (2 \times 3 \times 3 \times 5)$$

The LCM is

$$2 \times 2 \times 3 \times 3 \times 5 \times 7 = 1260 \ (2 \times 3 \text{ is in both numbers})$$

What do you notice about the repeated factors and the HCF?

The HCF is $2 \times 3 = 6$

Negative numbers

You will be expected to perform calculations like these without a calculator.

$$8 - 11 = -3 \qquad -5 + 1 = -4$$

$$7 + (-2) = 5 \qquad 7 - (-2) = 9$$

and

$$-7 \times -3 = 21 \qquad -7 \times 3 = -21$$

$$7 \times -3 = -21$$

Example

Copy and complete (indicated in red) this table:

Positive integer e.g. 12	÷ ÷	Positive integer 3	= =	Positive integer 4
Negative integer e.g. –12	÷ ÷	Positive integer 3	= =	Negative integer –4
Positive integer e.g. 12	÷ ÷	Negative integer –3	= =	Negative integer –4
Negative integer e.g. –12	÷ ÷	Negative integer –3	= =	Positive integer 4

PROGRESS TEST

1 Complete these addition, subtraction and multiplication grids.

+	3	–1	0	–	6	–5	1	×	5	–3	3
–6	–3	–7	–6	–3	–9	2	–4	–7	–35	21	–21
5	8	4	5	9	3	14	8	5	25	–15	15
–1	2	–2	–1	–10	–16	–5	–11	–4	–20	12	–12

2 Find the prime factorisation of the greatest 3-digit number.

3 Find the LCM of 72, 240 and 196. Leave your answer in index form.

4 Find the HCF of 72 and 126.

1

+	3	–1	0	–	6	–5	1	×	5	–3	3
–6	–3	–7	–6	–3	–9	2	–4	–7	–35	21	–21
5	8	4	5	9	3	14	8	5	25	–15	15
–1	2	–2	–1	–10	–16	–5	–11	–4	–20	12	–12

2 $999 = 3^3 \times 37$

3 $2^4 \times 3^2 \times 5 \times 7^2$

4 $2 \times 3 \times 3 = 18$

1.2 Powers and roots

LEARNING SUMMARY

After studying this section, you will be able to:

- ***find square roots and cube roots***
- ***use and understand the laws of indices***
- ***use and understand inverse operations***
- ***use standard index form***

Square roots and cube roots

KEY POINT

The square root of 9 is 3, because $9 = 3 \times 3$.
We use the symbol $\sqrt{\ }$ and write $\sqrt{9} = 3$.

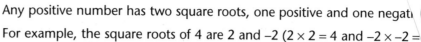

Any positive number has two square roots, one positive and one negative.
For example, the square roots of 4 are 2 and –2 ($2 \times 2 = 4$ and $-2 \times -2 = 4$).

> **KEY POINT**
>
> **The cube root of 8 is 2, because $8 = 2 \times 2 \times 2$. We use the symbol $\sqrt[3]{}$ and write $\sqrt[3]{8} = 2$.**

For example, $\sqrt[3]{64} = 4$ ($4 \times 4 \times 4 = 64$).

You can use your calculator to find square roots and cube roots. The keys may be labelled $\sqrt{}$ and $\sqrt[3]{}$ – get to know your own calculator by checking these.

$$\sqrt{169} = 13 \qquad \sqrt[3]{27} = 3 \qquad \sqrt{1.44} = 1.2 \qquad \sqrt[3]{15.625} = 2.5$$

Laws of indices

An **index** is a number that tells us how many times we must multiply another number by itself. For instance, 2^3 means $2 \times 2 \times 2 = 8$; we say 'three raised to the power of two'. The plural of index is **indices**.

For example, $2 \times 2 \times 2 \times 2 = 2^4$

$3 \times 3 \times 3 \times 3 \times 3 = 3^5$

In the expression 3^5, the **index** is 5 and the number 3 is called the **base**. Your calculator will probably have a button to evaluate powers of numbers. It may be marked x^y, $x \wedge$ or $x^\wedge y$. Check this, and then use your calculator to verify that $5^4 = 625$ and $13^7 = 62\,748\,517$.

To manipulate expressions involving indices we use rules, sometimes known as the **laws of indices**.

> **KEY POINT**
>
> **The index law for multiplication**
>
> **When expressions with the same base are multiplied, the indices are added:**
>
> $a^p \times a^q = a^{p+q}$
>
> **so $2^5 \times 2^3 = 2^8$**

Example

(a) Simplify $2 \times 2^2 \times 2^3$

$$= 2^{1+2+3} \quad (2 = 2^1)$$

$$= 2^6$$

(b) Simplify $(2^3)^4$

$$= 2^3 \times 2^3 \times 2^3 \times 2^3$$

$$= 2^{12}$$

> **KEY POINT**
>
> **The index law for division**
>
> When expressions with the same base are divided, the indices are subtracted:
>
> $a^p \div a^q = a^{p-q}$
>
> so $2^5 \div 2^3 = 2^2$

Remember you can only add the indices if numbers have the same base.

Example

Simplify $\dfrac{2^3 \times 3^4 \times 2^4}{2^2 \times 3^2}$

$= \dfrac{2^7 \times 3^4}{2^2 \times 3^2} = (2^7 \times 3^4) \div (2^2 \times 3^2) = 2^5 \times 3^2$

> **KEY POINT**
>
> When dealing with brackets, the index law for multiplication such as $(a^p)^q$ gives the result $a^{p \times q} = a^{pq}$. For example, $(3^5)^2 = 3^5 \times 3^5 = 3^{10}$ ($3^{5 \times 2} = 3^{10}$).

> **KEY POINT**
>
> **First special case of the index laws**
>
> Using the multiplication law $a^p \times a^0 = a^{p+0} = a^p$
>
> So a^0 must be equal to 1 – we have not given a value to a.
>
> This means that any number to the power zero is 1.

> **KEY POINT**
>
> **Second special case of the index laws**
>
> From the division law $1 \div a^p = a^{0-p} = a^{-p}$ (remember $a^0 = 1$)
>
> So $1 \div a^p = \dfrac{1}{a^p} = a^{-p}$
>
> This means, for example, that $4^{-2} = \dfrac{1}{4^2} = \dfrac{1}{16}$.

> **KEY POINT**
>
> **Third special case of the index laws**
>
> Using the multiplication law: $x^{\frac{1}{2}} \times x^{\frac{1}{2}} = x^{\left(\frac{1}{2}+\frac{1}{2}\right)} = x^1 = x$
>
> This shows that $x^{\frac{1}{2}} = \sqrt{x}$
>
> The general law is $\sqrt[n]{x} = x^{\frac{1}{n}}$.

Inverse operations

The **inverse** of a function or a transformation is the function or transformation that 'undoes' it.

The **inverse** of taking a reciprocal is taking the reciprocal again.

The inverse of addition is subtraction.

The inverse of a clockwise rotation is an anticlockwise rotation.

The inverse of $\sqrt{x}$ is x^2 (since $(\sqrt{x})^2 = x$.

The inverse of $\sqrt[n]{x}$ is x^n (since $(\sqrt[n]{x})^n = x$ (remember $\sqrt[n]{x} = x^{\frac{1}{n}}$ and $x^{\frac{1}{n} \times n} = x^1 = x$).

> **KEY POINT**
>
> The operation that reverses what has been done is called an inverse operation.

Standard index form

When a number is written in the form $a \times 10^b$, where $1 \leq a < 10$, then it is said to be in **standard** or **standard index form**.

For example,

$$200 = 2 \times 10^2 \quad 3000 = 3 \times 10^3 \quad 0.6 = 6 \times 10^{-1} \quad 0.047 = 4.7 \times 10^{-2}$$

$$312 = 3.12 \times 10^2 \quad 85 = 8.5 \times 10^1 \quad 0.000\,83 = 8.3 \times 10^{-4}$$

$$334\,682 = 3.346\,82 \times 10^5$$

but $1230 = 12.3 \times 10^3$ is **not** in standard form (because 12.3 is not between 1 and 10) and $1230 = 0.123 \times 10^4$ is not in standard form (because 0.123 is not between 1 and 10).

Find out how to use your calculator with numbers in standard index form – how to enter figures and how to read them and also how to use brackets and the power key to calculate, for example, $10^{\frac{1}{3}}$.

You can only add numbers in standard form when they have the same power of ten.

Examples

Using a calculator:

(a) Calculate to 3 s.f. (see page 32) $\dfrac{7.56 \times 10^8}{2.51 \times 10^3}$ and give your answer in standard form:

$$= 3.01 \times 10^5$$

(b) Calculate to 3 s.f. $\dfrac{7.89 \times 10^3}{\sqrt{6.44 \times 10^8}}$ and give your answer in standard form:

$$= 3.11 \times 10^{-1}$$

Without using a calculator:

(c) Work out $7.3 \times 10^4 + 6.1 \times 10^3$ and give your answer in standard form:

$$= 7.91 \times 10^4$$

(d) Work out $\dfrac{6 \times 10^4}{1.5 \times 10^2}$ and give your answer in standard form:

$$= 4 \times 10^2$$

1 Write as ordinary numbers:

(a) 3^4 (b) 1.07×10^4 (c) 2^{-3} (d) $4^{\frac{1}{2}}$ (e) $(2^2)^3$ (f) 8^0 (g) $\sqrt{100}$ (h) $^3\sqrt{1000}$

2 Simplify, if possible, leaving your answer in index form:

(a) $\sqrt{21}$ (b) $\frac{1}{4^2}$ (c) $5^6 \div 5^{10}$ (d) $3^2 \div 3^{-4}$ (e) $11^{-4} \div 11^0$ (f) $2^{-5} \div 2^{-1}$

(g) $3^2 \times 2^3$ (h) $4^2 \times 2^2$

3 Write in standard form, as simply as possible:

(a) $521\,000$ (b) $\frac{1}{25}$ (c) 1.001 (d) $\frac{1}{1000}$ (e) $(5 \times 10^3) \times (6 \times 10^2)$

(f) $(4 \times 10^2) \div (8 \times 10^3)$ (g) $\sqrt{9 \times 10^4}$ (h) $0.000\,101 \times 10^{-1}$

4 Work out the following, leaving your answer in standard index form, give your answers to 3 s.f.:

(a) $7.3 \times 10^3 \times 1.9 \times 10^2$ (b) $4.771 \times 10^3 \div (7.4 \times 10^3)$ (c) $\dfrac{\sqrt{4.5 \times 10^3}}{2.7 \times 10^{-3}}$

(d) $(2.5 \times 10^2)^3$

PROGRESS TEST

1.3 Fractions

After studying this section, you will be able to:

LEARNING SUMMARY

- **find and use equivalent fractions**
- **add and subtract fractions**
- **multiply fractions**
- **divide fractions**

Equivalent fractions

KEY POINT

Equivalent fractions are fractions that have the same value:

the fractions $\dfrac{1}{4}$, $\dfrac{2}{8}$, $\dfrac{4}{16}$, $\dfrac{8}{32}$ have the same value.

$$\frac{1}{4} \qquad \frac{2}{8} \qquad \frac{4}{16} \qquad \frac{8}{32}$$

Equivalent fractions are generated by multiplying (or dividing) the top (numerator) and the bottom (denominator) of a fraction by the same number.

For example, $\dfrac{20}{30} = \dfrac{4 \times 20}{4 \times 30} = \dfrac{80}{120} = \dfrac{80 \div 10}{120 \div 10} = \dfrac{8}{12}$.

Examples

(a) Write these fractions in order, smallest first.

$$\frac{5}{6}, \frac{17}{20}, \frac{4}{5}, \frac{2}{3}$$

We first need to find the **common denominator**; this must have factors 3, 5, 6 and 20. One way is to multiply these numbers together. This always works, but produces a very large number. (Remember fractions are easily compared if they have the same denominator.)

A more efficient way is to find the LCM of 3, 5, 6 and 20.

3 and 5 are prime so have no common factors (apart from 1), but $6 = 2 \times 3$ and $20 = 2 \times 2 \times 5$, so the LCM is $2 \times 2 \times 3 \times 5$.

$$\frac{5}{6} = \frac{5 \times 2 \times 5}{5 \times 2 \times 2 \times 3} = \frac{50}{60}, \qquad \frac{17}{20} = \frac{5 \times 17}{3 \times 20} = \frac{51}{60}$$

$$\frac{4}{5} = \frac{4 \times 3 \times 4}{4 \times 3 \times 5} = \frac{48}{60}, \qquad \frac{2}{3} = \frac{5 \times 4 \times 2}{5 \times 4 \times 3} = \frac{40}{60}$$

So the order is $\dfrac{2}{3}, \dfrac{4}{5}, \dfrac{5}{6}, \dfrac{17}{20}$.

(b) Write as simply as possible: $\dfrac{252}{336}$

> This process is called cancelling. It can be done in easy stages.

$$\frac{252}{336} = \frac{252 \div 2}{336 \div 2} = \frac{126}{168} = \frac{126 \div 2}{168 \div 2} = \frac{63}{84} = \frac{63 \div 7}{84 \div 7} = \frac{9}{12} = \frac{9 \div 3}{12 \div 3} = \frac{3}{4}$$

Addition and subtraction of fractions

Adding, subtracting or comparing fractions with the same denominator (bottom number) is straightforward, for example: $\dfrac{3}{7} + \dfrac{2}{7} = \dfrac{5}{7}$: put in words *'three-sevenths and two-sevenths added gives five-sevenths'*.

> The LCM of 8 and 3 is 24.

> The LCM of 3 and 12 is 12.

KEY POINT

To add, subtract or compare fractions, arrange for them to have the same denominator (and to keep their value) using equivalent fractions. This common denominator is the LCM of the original denominators.

For example, $\dfrac{3}{8} + \dfrac{1}{3} = \dfrac{3 \times 3}{3 \times 8} + \dfrac{8 \times 1}{8 \times 3} = \dfrac{9}{24} + \dfrac{8}{24} = \dfrac{17}{24}$

and $\dfrac{2}{3} - \dfrac{1}{12} = \dfrac{4 \times 2}{4 \times 3} - \dfrac{1}{12} = \dfrac{8}{12} - \dfrac{1}{12} = \dfrac{7}{12}$

When adding fractions never add the numerators together and the denominators together – remember, for example, $\frac{1}{2} + \frac{1}{2}$ is not $\frac{2}{4}$.

Multiplication of fractions

 KEY POINT To multiply two fractions, multiply the numerators and multiply the denominators.

Examples

(a) Find $\dfrac{3}{4} \times \dfrac{5}{7}$

$$\dfrac{3}{4} \times \dfrac{5}{7} = \dfrac{3 \times 5}{4 \times 7} = \dfrac{15}{28}$$

(b) Work out $\dfrac{5}{6} \times \dfrac{2}{3}$

$$\dfrac{5}{6} \times \dfrac{2}{3} = \dfrac{10}{18} = \dfrac{5}{9}$$

The 2 could have been cancelled.

$$\dfrac{5}{\overset{2}{\underset{3}{6}}} \times \dfrac{\overset{1}{2}}{3} = \dfrac{5}{9}.$$

(c) Work out $\dfrac{2}{7} \times 11$

$$\dfrac{2}{7} \times 11 = \dfrac{2}{7} \times \dfrac{11}{1} = \dfrac{22}{7} = 3\dfrac{1}{7}$$

Any integer can be written with 1 as its denominator, i.e. $n = \dfrac{n}{1}.$

Division of fractions

Look carefully at this division: $2 \div \dfrac{1}{3}$. It is asking 'How many thirds are there in two?'

The answer is $2 \times 3 = 6$. Here are two more examples:

$$3 \div \dfrac{1}{4} = 3 \times 4 = 12 \qquad 4 \div \dfrac{1}{2} = 4 \times 2 = 8$$

This suggests that for two integers m and n: $\ n \div \dfrac{1}{m} = n \times m$

Here's another, more complicated, division: $\dfrac{5}{8} \div \dfrac{2}{3}$

From the above it looks like the answer might be $\dfrac{5}{8} \times \dfrac{3}{2} = \dfrac{15}{16}$

We can check this because if $\dfrac{5}{8} \div \dfrac{2}{3} = \dfrac{15}{16}$ then $\dfrac{5}{8} = \dfrac{2}{3} \times \dfrac{15}{16}$ which is true!

(If you are not convinced look at this: $20 \div 4 = 5$, so $20 = 4 \times 5$.)

Swapping over a fraction's numerator and denominator is called taking its **reciprocal**:

the reciprocal of $\dfrac{2}{3}$ is $\dfrac{3}{2}$, the reciprocal of $\dfrac{3}{2}$ is $\dfrac{2}{3}$, the reciprocal of $\dfrac{1}{2}$ is $\dfrac{2}{1} = 2$.

KEY POINT To divide by a fraction, multiply by its reciprocal.

Example

$$\frac{2}{7} \div \frac{2}{3}$$

$$\frac{2}{7} \div \frac{2}{3} = \frac{2}{7} \times \frac{3}{2} = \frac{6}{14} = \frac{3}{7}$$

Mixed fractions

An **improper fraction** has a numerator that is greater than or equal to its denominator, for example $\frac{7}{4}$. An improper, or 'top-heavy', fraction can be

written as a **mixed fraction**: $\frac{7}{4} = 1\frac{3}{4}$.

Mixed fractions sometimes occur in fraction calculations. In most cases it is best to change these into improper fractions before performing the calculation.

Examples

(a) Work out $18 \div \frac{3}{4}$

> 3 has been cancelled to save on work.

$$18 \div \frac{3}{4} = \frac{18}{1} \times \frac{4}{3} = \frac{24}{1} = 24$$

(b) Work out $\frac{3}{4}$ of $1\frac{1}{2}$

> 'of' can be replaced by 'multiply'.

$$\frac{3}{4} \times 1\frac{1}{2} = \frac{3}{4} \times \frac{3}{2} = \frac{9}{8} = 1\frac{1}{8}$$

(c) Calculate $1\frac{2}{3} + \frac{3}{5}$

> Changing $1\frac{2}{3}$ into an improper fraction. LCM of 3 and 5 is 15.

$$1\frac{2}{3} + \frac{3}{5} = \frac{5}{3} + \frac{3}{5} = \frac{25}{15} + \frac{9}{15} = \frac{34}{15} = 2\frac{4}{15}$$

(d) Work out $1\frac{5}{8} - \frac{2}{3}$

$$1\frac{5}{8} - \frac{2}{3} = \frac{13}{8} - \frac{2}{3} = \frac{39}{24} - \frac{16}{24} = \frac{23}{24}$$

> When adding fractions involving mixed numbers it is sometimes easier to add the whole numbers and add the fractions separately – this does not always help when subtracting fractions.

(e) Work out $3\frac{1}{2} + 2\frac{4}{5}$

$$3\frac{1}{2} + 2\frac{4}{5} = 5 + \frac{1}{2} + \frac{4}{5} = 5 + \frac{13}{10} = 6\frac{3}{10}$$

Number

 PROGRESS TEST

1 Find the simplest fractions equivalent to these:

(a) $\frac{652}{978}$ (b) $\frac{432}{288}$ (c) $\frac{147}{343}$ (d) $\frac{978}{652}$

2 Change these improper fractions into mixed fractions:

(a) $\frac{19}{3}$ (b) $\frac{14}{6}$ (c) $\frac{17}{5}$ (d) $\frac{63}{10}$

3 Change these mixed fractions to improper fractions:

(a) $3\frac{1}{4}$ (b) $3\frac{3}{8}$ (c) $2\frac{2}{7}$ (d) $1\frac{2}{5}$

4 Work these out:

(a) $1\frac{3}{4} - \frac{1}{3}$ (b) $1\frac{1}{3} - \frac{2}{3}$ (c) $2\frac{3}{4} + 2\frac{2}{3}$ (d) $1\frac{3}{10} + \frac{1}{5} + \frac{2}{3}$

5 Work these out:

(a) $\frac{3}{4} \times \frac{4}{5}$ (b) $\frac{3}{4} \times \frac{7}{8}$ (c) $2\frac{1}{4} \times 1\frac{1}{3}$ (d) $20 \div \frac{4}{5}$ (e) $1\frac{1}{3} \div \frac{2}{5}$

(f) $\left(1\frac{1}{3}\right)^2$ (g) $\frac{2}{3}$ of $13\frac{1}{2}$ (h) $2\frac{1}{4} \div 1\frac{1}{3}$

1 (a) $\frac{2}{3}$ (b) $\frac{3}{2}$ (c) $\frac{3}{7}$ (d) $\frac{3}{2}$

2 (a) $6\frac{1}{3}$ (b) $2\frac{1}{3}$ (c) $3\frac{2}{5}$ (d) $6\frac{3}{10}$

3 (a) $\frac{13}{4}$ (b) $\frac{27}{8}$ (c) $\frac{16}{7}$ (d) $\frac{7}{5}$

4 (a) $1\frac{5}{12}$ (b) $\frac{2}{3}$ (c) $5\frac{5}{12}$ (d) $2\frac{1}{6}$

5 (a) $\frac{3}{5}$ (b) $\frac{21}{32}$ (c) 3 (d) 25 (e) $3\frac{1}{3}$ (f) $1\frac{7}{9}$ (g) 9 (h) $1\frac{11}{16}$

1.4 Decimals

 LEARNING SUMMARY

After studying this section, you will be able to:

- *change a fraction into a decimal*
- *change a decimal into a fraction*
- *recognise recurring decimals*

Decimals and fractions

KEY POINT To change a fraction into a decimal, divide the numerator by the denominator.

Examples

(a) $\frac{3}{5} = 3 \div 5 = 0.6$

(b) $\frac{2}{3} = 2 \div 3 = 0.666\,666\,6 \ldots$

(c) $\frac{3}{11} = 3 \div 11 = 0.272\,727\,27 \ldots$

The digits in the decimals in (b) and (c) have a repeating pattern and go on forever. Decimals that do this are called **recurring decimals**.

22

> **KEY POINT**
>
> All fractions give either **terminating** or **recurring** decimals. All terminating decimals are rational numbers (see page 35), but not all rational numbers have terminating decimal expansions.

Example

Which of these fractions give recurring decimals? $\dfrac{7}{8}, \dfrac{2}{21}, \dfrac{11}{32}, \dfrac{5}{37}$

$\dfrac{7}{8} = 0.875$ (terminating) $\dfrac{2}{21} = 0.095\,230\,952\,30\ldots$ (recurring)

$\dfrac{11}{32} = 0.343\,75$ (terminating) $\dfrac{5}{37} = 0.135\,135\,135\ldots$ (recurring)

There are two ways of showing the repeating digits of a recurring decimal: using either a dot or a line above the number or numbers that make the pattern:

$\dfrac{5}{27} = 5 \div 27 = 0.185\,185\,185\ldots$

which can be written as $0.\dot{1}8\dot{5}$ or $0.\overline{185}$

> **KEY POINT**
>
> To change a decimal into a fraction, use place value: so 0.24 is represented by $\dfrac{2}{10} + \dfrac{4}{100} = \dfrac{24}{100} = \dfrac{6}{25}$ in simplest terms.

Example

Find fractions equivalent to these decimals:

0.3   0.125   0.031 25 $0.\dot{3}$ $0.\dot{3}\dot{9}$ $0.\dot{3}1\dot{9}$

$0.3 = \dfrac{3}{10}$

$0.125 = \dfrac{125}{1000} = \dfrac{125 \div 25}{1000 \div 25} = \dfrac{5}{40} = \dfrac{1}{8}$

$0.031\,25 = \dfrac{3125}{100\,000} = \dfrac{3125 \div 5}{100\,000 \div 5} = \dfrac{625}{20\,000} = \dfrac{625 \div 25}{20\,000 \div 25} = \dfrac{1}{32}$

$0.\dot{3}$ needs a different approach, as do all recurring decimals

$10 \times 0.\dot{3} = 3.333\,333\,333\ldots$ (a)

$0.\dot{3} = 0.333\,333\,333\,333\ldots$ (b)

3.333 333 333 …
− 0.333 333 33 …
= 3.0

Check this result using your calculator.

(a) − (b) gives $10 \times 0.\dot{3} - 0.\dot{3} = 3$

so $9 \times 0.\dot{3} = 3$

which means that $0.\dot{3} = \dfrac{3}{9} = \dfrac{1}{3}$

$0.\dot{3}\dot{9}$ – a similar method can be used:

$100 \times 0.\dot{3}\dot{9} = 39.393\,939\,39\ldots$

$0.\dot{3}\dot{9} \qquad = 0.393\,939\,39\ldots$

Subtracting these two expressions gives:

$99 \times 0.\dot{3}\dot{9} = 39$

so $0.\dot{3}\dot{9} = \dfrac{39}{99} = \dfrac{13}{33}$

Similarly for $0.\dot{3}1\dot{9}$

$1000 \times 0.\dot{3}1\dot{9} = 319.319\,319\,319\ldots$

$ 0.\dot{3}1\dot{9} = 0.319\,319\,319\ldots$

giving

$ 999 \times 0.\dot{3}1\dot{9} = 319$

so

$$0.\dot{3}1\dot{9} = \frac{319}{999}$$

PROGRESS TEST

1 Write as decimals:

(a) $\dfrac{9}{16}$ (b) $\dfrac{7}{25}$ (c) $\dfrac{5}{44}$ (d) $\dfrac{1}{27}$

2 Write the following as fractions in their lowest terms:

(a) 0.45 (b) $0.4\dot{0}\dot{5}$ (c) $0.\dot{0}\dot{3}$ (d) $0.\dot{4}$

1 (a) 0.5625 (b) 0.28 (c) 0.1136 (d) 0.037

2 (a) $\dfrac{45}{100} = \dfrac{9}{20}$ (b) $\dfrac{405}{999} = \dfrac{15}{37}$ (c) $\dfrac{3}{90} = \dfrac{1}{30}$ (d) $\dfrac{4}{9}$

1.5 Percentages

LEARNING SUMMARY

After studying this section, you will be able to:

- *change percentages to fractions and vice versa*
- *change percentages to decimals and vice versa*
- *express one quantity as a percentage of another*
- *find a percentage of a quantity*
- *find a percentage increase or decrease*
- *use reversed percentages*

Percentages and fractions

Percentages are a way of expressing a fraction as a number of parts out of 100. For instance, 25 per cent (25%) means $\dfrac{25}{100}$, which can be simplified to $\dfrac{1}{4}$.

In general $n\%$ is equivalent to the fraction $\dfrac{n}{100}$.

KEY POINT

To change a percentage to a fraction, write it over 100 and cancel any common factors.

Example

Write as fractions:

45%, 33%, 145%

$45\% = \dfrac{45}{100} = \dfrac{9}{20}$ ← **Cancelling by 5**

$33\% = \dfrac{33}{100}$ ← **Care! 33% is not $\frac{1}{3}$, which is $33\frac{1}{3}$%**

$145\% = \dfrac{145}{100} = \dfrac{29}{20} = 1\dfrac{9}{20}$ ← **The result is larger than 1 as the percentage is larger than 100**

KEY POINT To change a fraction to a percentage, multiply by 100%.

Example

Change to percentages:

$\dfrac{2}{5}, \dfrac{42}{700}, \dfrac{2}{3}, 1\dfrac{1}{2}, 1\dfrac{n}{100}$

$\dfrac{2}{5} \times 100\% = \dfrac{2 \times 100}{5}\% = 2 \times 20\% = 40\%$

$\dfrac{42}{700} \times 100\% = \dfrac{42 \times 100}{700}\% = \dfrac{42}{7}\% \times = 6\%$

$\dfrac{2}{3} \times 100\% = \dfrac{2 \times 100}{3}\% = \dfrac{200}{3}\% = 66\dfrac{2}{3}\%$

$1\dfrac{1}{2} \times 100\% = 150\%$

$1\dfrac{n}{100} = 1\dfrac{n}{100} \times 100\% = (100 + n)\%$

Percentages and decimals

KEY POINT To change a percentage to a decimal, write it over 100 and divide it out.

Example

Write as decimals:

37%   7%   250%   0.5%

$\dfrac{37}{100} = 0.37, \quad \dfrac{7}{100} = 0.07, \quad \dfrac{250}{100} = 2.5, \quad \dfrac{0.5}{100} = 0.005$

KEY POINT To change a decimal to a percentage, multiply by 100%.

Example

Write as percentages:

0.6   0.72   0.009   0.000 67

0.6 = 0.6 × 100% = 60%   0.72 = 0.72 × 100% = 72%

0.009 = 0.009 × 100% = 0.9%   0.000 67 = 0.000 67 × 100% = 0.067%

One quantity as a percentage of another

 KEY POINT   To find one quantity as a percentage of another, divide the first by the second and multiply by 100%.

Examples

(a) What is 45 as a percentage of 150?

$$\text{Percentage} = \frac{45}{150} \times 100\% = \frac{3}{10} \times 100\% = 30\%$$

> Cancelling before multiplying eases the arithmetic if not using a calculator

(b) What is 39 out of 72 as a percentage?

Give your answer correct to one decimal place.

$$\text{Percentage} = \frac{39}{72} \times 100\% = 0.541\,66 \ldots \times 100\% = 54.2\% \text{ (to 1 d.p.)}$$

(c) The price of a small flat increases from £125 000 to £150 000.

What is the percentage change?

> The percentage change of a quantity is always the percentage of the *original* value.

The change is £25 000.

So the percentage change from £125 000 is $\dfrac{25\,000}{125\,000} \times 100\%$

$$= \frac{25 \times 100}{125} = 20\%$$

Finding a percentage of a quantity

 KEY POINT   To find a percentage of a quantity, multiply by the percentage and then divide by 100.

Examples

(a) Find 8% of 42.

$$\frac{8 \times 42}{100} = \frac{336}{100} = 3.36$$

(b) What is 17.5% of £380?

$$\frac{17.5 \times 380}{100} = \frac{6650}{100} = £66.50$$

Finding a percentage increase or decrease

Example

A 'collectable' toy cost £50 last year. It has increased 15% in cost.

What is the new cost?

15% of £50 is $£\dfrac{15}{100} \times 50$.

So the new cost is $£50 + £\dfrac{15}{100} \times £50 = £50 + £7.50$ (old price + 15% increase)

$$= £57.50$$

> **Quicker than adding the increase.**

$50 + \dfrac{15}{100} \times £50$ is the same as multiplying 50 by $1 + \dfrac{15}{100}$ (or 1.15)

> **KEY POINT**
>
> To find the result of a percentage increase, multiply by
> (1 + the percentage divided by 100) – this is sometimes called the
> multiplier.

Examples

(a) Increase £84 by 23%.

To increase by 23% multiply by $1 + \dfrac{23}{100}$ (1.23)

$84 \times 1.23 = £103.32$

(b) The height of a tree increased by 5%.

It was 8 m tall. How tall is it now?

To increase by 5% multiply by $1 + \dfrac{5}{100}$ (1.05)

$8 \times 1.05 = 8.4$ m

(c) Write down the multiplier for a percentage increase of:

   (i)   2.5%   Answer: 1.025   (ii) 150%   Answer: 2.5

   (iii) 100%   Answer: 2   (iv) 0.1%   Answer: 1.001

A similar method can be used when calculating the results of percentage decreases.

Example

A 'collectable' toy cost £50 last year. It decreased by 15% in cost.

What is the new cost?

15% of £50 is $£\dfrac{15}{100} \times 50$.

So the new cost is $£50 - £\dfrac{15}{100} \times 50 = £50 - £7.50$ (old price – 15% decrease)

$$= £42.50$$

$50 - \dfrac{15}{100} \times 50$ is the same as multiplying 50 by $1 - \dfrac{15}{100}$ (or 0.85).

> **KEY POINT**
>
> To find the result of a percentage decrease, multiply by
> (1 – the percentage divided by 100): the multiplier.

Examples

(a) Write down the multiplier for a percentage decrease of:

   (i)   20%   Answer: 0.8   (ii) 5%   Answer: 0.95

   (iii) 2.5% Answer: 0.975   (iv) 0.1%   Answer: 0.999

(b) A pair of jeans is priced at £27.

   During a sale it was reduced by 30%. How much do the jeans cost now?

   To decease by 30% multiply by 0.7. ⟵ $1 - \dfrac{30}{100}$

$$0.7 \times 27 = £18.90$$

(c) I bought a car for £12 500.

   A year later its value has fallen by 19%.

   What is its value now?

   To decrease by 19% multiply by 0.81. ⟵ $1 - \dfrac{19}{100}$

$$12\,500 \times 0.81 = £10\,125$$

Reversed percentages

Examples

> The answer is not simply £140 reduced by 12%! Try it and see.

(a) A bike costs £140 after a 12% increase in price.

   How much did it cost before the increase?

   To find an increase of 12% we use a multiplier of 1.12.

   So old price × 1.12 = £140 ⟵ **We have an equation**

   So old price = £140 ÷ 1.12

              = £125 ⟵ **Dividing both sides by 1.12**

(b) In a sale with '20% off everything' a skirt cost £24. How much was it before the sale?

   To find a decrease of 20% a multiplier of 0.8 is needed:

> You should be especially careful when doing reversed percentages – check by working through the question with your answer – does it fit?

   old price × 0.8 = £24 ⟵ **An equation in 'old price'**

   so old price = £24 ÷ 0.8

             = £30

Check your 'reversed percentage' calculations by increasing or decreasing your answers by the given percentage to see that it fits the question.

> **KEY POINT**
> To find the value before a percentage **increase**, divide by
> (1 + the percentage divided by 100).

> **KEY POINT**
> To find the value before a percentage **decrease**, divide by
> (1 – the percentage divided by 100).

Examples

(a) A TV is reduced by 20% in a sale. Its sale price was £176.

How much did it cost before the sale?

$$\text{old price} \times 0.8 = £176$$
$$\text{so old price} = £176 \div 0.8$$
$$= £220$$

(b) Over the last year petrol has increased in price by 7.5%.

A litre of petrol is now 102p.

How much was a litre of petrol a year ago?

$$\text{old price} \times 1.075 = 102\text{p}$$
$$\text{so old price} = 102\text{p} \div 1.075$$
$$\text{which is 95p to the nearest p.}$$

PROGRESS TEST

1 Change to fractions:
(a) 15% (b) 125% (c) 0.5% (d) 65%
2 Change to decimals:
(a) 21% (b) $17\frac{1}{2}$% (c) 205% (d) $\frac{1}{4}$%
3 Change to percentages:
(a) $\frac{3}{16}$ (b) $2\frac{1}{4}$ (c) 0.012 (d) 1.05
4 Find:
(a) 25 out of 40 as a percentage (b) 6 as a percentage of 30
(c) 0.7 as a percentage of 20 (d) 45 as a percentage of 1500
5 Find:
(a) 18% of 680 (b) 7.5% of 87 (c) $\frac{3}{4}$% of 60 (d) 0.3% of 15.5
6 Increase:
(a) 15 by 5% (b) 2.1 by 30% (c) 9.2 by 5.5% (d) 144 by 7.5%
7 Decrease:
(a) 321 by 15% (b) 752 by 1% (c) 40.8 by 2.5% (d) 500 by 85%
8 Find the amount:
(a) that was increased by 5% to give £116.55
(b) that was increased by 7.5% to give £43
(c) that was decreased by 5% to give £55.10
(d) that was decreased by 35% to give £195

8 (a) £111 (b) £40 (c) £58 (d) £300
7 (a) 272.85 (b) 744.48 (c) 39.78 (d) 75
6 (a) 15.75 (b) 2.73 (c) 9.706 (d) 154.8
5 (a) 122.4 (b) 6.525 (c) 0.45 (d) 0.0465
4 (a) 62.5% (b) 20% (c) 3.5% (d) 3%
3 (a) 18.75% (b) 225% (c) 1.2% (d) 105%
2 (a) 0.21 (b) 0.175 (c) 2.05 (d) 0.0025
1 (a) $\frac{3}{20}$ (b) $1\frac{1}{4}$ (c) $\frac{1}{200}$ (d) $\frac{13}{20}$

1.6 Ratio

 LEARNING SUMMARY

After studying this section, you will be able to:

- simplify ratios
- divide in a ratio

Simplifying ratios

KEY POINT

A ratio compares the relative sizes of two or more quantities.

For example, the ratio of crosses (×) to noughts (0) in each of these boxes is 2 to 1 or 2:1.

| XX0 | XXXX00 | XXXXXX000 |

- Ratios can be simplified by dividing each part of the ratio by the same number.
- A ratio is unchanged if each part is multiplied (or divided) by the same number.
- To compare quantities they must be in the same units, for example: 4 kg to 500 g is 4000:500, which simplifies to 8:1, not 4:500.

> **Example**
> Write these ratios in their simplest form:
> (a) 4:12   (b) 18:81   (c) 36:4   (d) 20:70
>   4:12 = 1:3   18:81 = 2:9   36:4 = 9:1   20:70 = 2:7

Dividing in a ratio

Share £200 in the ratio 2:3.

There are 2 + 3 = 5 equal parts to be shared.

These 5 parts are to share £200, so 1 part is 'worth' 200 ÷ 5 = £40.

The money should be shared 2 × £40 and 3 × £40 or £80 and £120.

or

£200 is to be divided in the ratio 2:3, there are 2 + 3 = 5 parts.

So the first share is therefore $\frac{2}{5}$ of the 200, which is £80 and so on.

> **KEY POINT**
>
> To divide a quantity in a particular ratio, first add the parts of the ratio (this gives the total number of parts to be divided or shared). Then divide the quantity to be shared by this total. You now know the value of a single share. Now multiply each part of the ratio by this amount to find out how the quantity is shared out.

It's a good idea to check the answers to make sure that the answers add to what is to be shared out.

Examples

(a) Four friends share a prize of £7000 in the ratio 4 : 5 : 3 : 2.

How much should each person receive?

There are 4 + 5 + 3 + 2 shares, so each share is worth

£7000 ÷ 14 = £500

The first person gets 4 shares, which is worth 4 × £500 = £2000.

The second person gets 5 shares, which is worth 5 × £500 = £2500.

The third person and fourth people get £1500 and £1000, respectively.

(Check: £2000 + £2500 + £1500 + £1000 = £7000.)

(b) The ratio of two brothers' weights is 3 : 4.

The lighter brother weighs 57 kg.

How much does the other brother weigh?

57 kg is 3 parts, so 1 part is 57 ÷ 3 = 19 kg

The heavier brother weighs 4 parts which is 4 × 19 = 76 kg.

PROGRESS TEST

1 Simplify these ratios:

(a) 80 : 100   (b) 3 km : 20 m   (c) $3\frac{1}{2}$: 2   (d) 18 : 27 : 72

(e) 1 hour : 12 minutes

2 Share in the given ratio:

(a) £720 in the ratio 4 : 5   (b) 180 kg in the ratio 1 : 2 : 3 : 4

(c) £200 in the ratio 1 : 4 : 3   (d) £600 in the ratio 4 : 5 : 6

2 (a) £320, £400   (b) 18 kg, 36 kg, 54 kg, 72 kg   (c) £25, £100, £75   (d) £160, £200, £240
1 (a) 4 : 5   (b) 150 : 1   (c) 7 : 4   (d) 2 : 3 : 8   (e) 5 : 1

1.7 Mental methods

 LEARNING SUMMARY

After studying this section, you will be able to:

- *recall number facts*
- *round numbers*
- *make estimates*

Number facts

KEY POINT

Make sure you know:

- All integer squares from 2^2 (4) up to 15^2 (225).
- The square roots of the perfect squares from 1 to 225.
- The integer cubes, 2^3 (8), 3^3 (27), 4^3 (64), 5^3 (125) and $10^3 = 1000$.
- The cube roots of 8, 27, 64, 125 and 1000.
- The multiplication tables up to 10×10.

Rounding numbers

 When you round, for example, to the nearest 100 do not discard the 0s – 3468 to the nearest 100 is 3500, not 35.

Examples

45 547 to the nearest hundred is 45 500 (45 547 has been rounded down)

45 576 to the nearest hundred is 45 600 (45 576 has been rounded up)

0.007 362 = 0.0074 correct to 4 decimal places (or d.p.)

$\qquad$ = 0.007 correct to 3 d.p.

$\qquad$ = 0.01 correct to 2 d.p.

If the first figure to be discarded is 5 or more, the previous figure is increased by 1 (or rounded up). Measurements written as 1.27, 1.23 or 1.20 are correct to 2 d.p., while 1.2 is correct to 1 d.p.

A second way of approximating a number is to use **significant figures** (s.f.). The first significant figure in any number is the non-zero figure with the highest place value.

In each of these numbers, the first significant figure is in **red**:

42 176 $\quad$ **2**.8997 $\quad$ 0.000 **7**438 $\quad$ 0.00**1** 607 4

For these numbers, the third significant figure is in **red**:

42 1**7**6 $\quad$ 2.8**9**97 $\quad$ 0.000 74**3** 8 $\quad$ 0.001 6**0**7 4

All these numbers have three significant figures:

1.06 $\quad$ 0.0000 10 6 $\quad$ 106 000 000

> **KEY POINT** The number of significant figures is found by ignoring leading zeros and counting the other figures including zeros.

This table shows some numbers rounded off to one, two and three significant figures.

Number	Rounded to 1 s.f.	Rounded to 2 s.f.	Rounded to 3 s.f.
45 481	50 000	45 000	45 500
8054	8000	8100	8050
0.7002	0.7	0.70	0.700
0.000 010 05	0.000 01	0.000 010	0.000 010 1

Making estimates

To estimate the result of a calculation, round or approximate each number so that you can work it out in your head. Rounding to one significant figure is often best.

Sometimes, especially when dividing, you may be able to round off a number to something more useful at 2 s.f. instead of 1 s.f.:

$57.3 \div 6.87 \approx 56 \div 7 = 8$

($\approx$ means approximately equal to).

Save the rounding until the end of your calculation – premature rounding can give wrong final answers.

> **Example**
>
> Estimate the answers to these calculations:
>
> (a) $601 \times (71.63 - 21.45)$
>
> (b) $(7.80 \times 10^4) \div (\sqrt{17})$
>
> (c) $\dfrac{36.87 \times 8.247}{15.689}$
>
> (d) $\dfrac{58.12}{48.2 \times 68.45}$
>
> (a) $601 \times (71.63 - 21.45) \approx 600 \times (70 - 20) = 30\,000$
>
> (b) $(7.80 \times 10^4) \div (\sqrt{17}) \approx 80\,000 \div 4 = 20\,000$
>
> (c) $\dfrac{36.87 \times 8.247}{15.689} \approx \dfrac{40 \times 8}{16} = 20$
>
> (d) $\dfrac{58.12}{48.2 \times 68.45} \approx \dfrac{60}{50 \times 70} = \dfrac{60}{3500} \approx \dfrac{60}{3600} = \dfrac{1}{60}$
>
> Checks for accuracy are very useful – but they are only rough. They ensure, for example, that your answer is not 10 times too big or 10 times too small.

1 State to the number of decimal places shown:
   (a) 1.05 (1 d.p.)   (b) 1.999 (2 d.p.)   (c) 16.439 71 (3 d.p.)
   (d) 45.004 31 (2 d.p.)   (e) 200.5050 (1 d.p.)   (f) 0.000 34 (1 d.p.)

2 State to the number of significant figures shown:
   (a) 48 460 (2 s.f.)   (b) 0.0348 (1 s.f.)   (c) 0.925 (1 s.f.)
   (d) 489.6 (3 s.f.)   (e) 1.008 (3 s.f.)   (f) 2.086 (2 s.f.)

3 Estimate the answers to these:
   (a) $\dfrac{38.98 \times \sqrt{101.9}}{1.9^2}$ (b) $\dfrac{1.55 \times 10^2 \times 0.917}{102.7 \times 0.0314}$ (c) $0.048 \div 0.099$

   (d) $\dfrac{12.7 + 34.9}{78.2 - 29.3}$ (e) $72.1 \times 3.225 \times 5.23$ (f) $\dfrac{4.9^2 \times 3.142}{\sqrt{24}}$

1 (a) 1.1 (b) 2.00 (c) 16.440 (d) 45.00 (e) 200.5 (f) 0.0
2 (a) 48 000 (b) 0.03 (c) 0.9 (d) 490 (e) 1.01 (f) 2.1
3 (a) 100 (b) 60 (c) 0.5 (d) 1 (e) 1100 (f) 15

1.8 Written methods

After studying this section, you will be able to:

● **use pencil and paper methods to perform simple calculations without a calculator**
● **calculate proportional changes**
● **recognise irrational numbers**
● **use surds in calculations**

Pencil and paper

You need to be able to perform calculations like these without the use of a calculator:

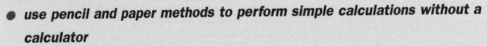

Write down the prime factors of 52 $[2^2 \times 13]$ Increase £15 by 20% [£18]

Simplify the fraction $\dfrac{65}{78}$ $\left[= \dfrac{5}{6}\right]$

Simplify the fraction $\dfrac{22 \times 144}{44 \times 3}$ $[= 24]$

Calculate: 137×67 $[= 9179]$

Calculate: $276 \div 12$ $[= 23]$

Calculate: $4 \times \sqrt{196}$ $[= 56]$

Calculate: $\dfrac{15}{64} \times \dfrac{16}{25}$ $\left[= \dfrac{3}{20}\right]$

Proportional change

> **KEY POINT**
> Proportional change is when a quantity is increased or deceased in a given ratio. This is achieved by using a multiplier.

> **First convert the ratio into $n:1$. The value of n gives the multiplier.**

For example,

To increase 20 in the ratio 3 : 2.

The ratio 3 : 2 = 1.5 : 1, giving 1.5 as the multiplier.

So the answer is $1.5 \times 20 = 30$.

> **You find the multiplier in the same way as with increasing.**

To decrease 15 in the ratio 2 : 3.

The ratio 2 : 3 = $\frac{2}{3}$: 1, so the multiplier is $\frac{2}{3}$.

> **It is best to leave fractions to the end. Answers are usually integers (at best) or terminating decimals.**

So the answer is $\frac{2}{3} \times 15 = 10$.

Irrational numbers

> **KEY POINT**
> A rational number is one that can be written as a fraction with numerator and denominator both integers.

> **KEY POINT**
> An irrational number is a number that cannot be written as a simple fraction. Irrational numbers have decimal expansions that neither terminate nor recur.

Examples of rational numbers are:

$3\ (=\frac{3}{1})$ $0.23\ (=\frac{23}{100})$ $0.\dot{3}0\dot{1}\ (=\frac{301}{999})$

Examples of irrational numbers are:

π $\sqrt{2}$ $2 + \sqrt{5}$ $\dfrac{\sqrt{3}}{2}$ $\dfrac{1}{\sqrt{2}}$

It can be proved that the square root of a prime number is always an irrational number.

Calculating with surds

 KEY POINT

A **surd** is a square root that cannot be evaluated exactly. Surds are therefore irrational numbers. The reason we leave them as surds is because in decimal form they would only be approximate, as they go on forever:

$\sqrt{25} = \pm 5$ is not a surd because it is exact, but $\sqrt{2} = 1.414\,213\,56\ldots$ is a surd.

There are two basic rules for manipulating surds:

If $\sqrt{x}$, $\sqrt{y}$ are surds, and a, b are 'ordinary' numbers

$$a\sqrt{x} \times b\sqrt{y} = ab\sqrt{xy} \quad \text{for example: } 6\sqrt{3} \times 3\sqrt{2} = 18\sqrt{6}$$

$$a\sqrt{x} \div b\sqrt{y} = \frac{a}{b}\sqrt{\frac{x}{y}} \quad \text{for example: } 10\sqrt{5} \div 2\sqrt{2} = \frac{10}{2}\sqrt{\frac{5}{2}} = 5\sqrt{\frac{5}{2}}$$

Examples

Simplify these expressions:

(a) $\dfrac{2\sqrt{32} \times 3\sqrt{2}}{\sqrt{8}}$

$= \dfrac{6\sqrt{32 \times 2}}{\sqrt{8}}$ ← **Simplify the product first**

$= 6\sqrt{\dfrac{64}{8}} = 6\sqrt{8} = 6 \times \sqrt{4 \times 2}$ ← **Then the division**

$= 6 \times \sqrt{4} \times \sqrt{2}$

$= 6 \times 2\sqrt{2}$

$= 12\sqrt{2}$

(b) $\dfrac{3\sqrt{5} \times 3\sqrt{6}}{\sqrt{30}}$

$= \dfrac{9\sqrt{30}}{\sqrt{30}} = 9$

(c) $\dfrac{7}{\sqrt{3}}$

$= \dfrac{\sqrt{3} \times 7}{\sqrt{3} \times \sqrt{3}} = \dfrac{7\sqrt{3}}{3}$ ← **Multiplying 'top' and 'bottom' by $\sqrt{3}$ removes the $\sqrt{3}$ in the denominator – a useful technique.**

Show that $\dfrac{7}{1 + 2\sqrt{5}}$ is an irrational number.

First look at $(1 + 2\sqrt{5}) \times (1 - 2\sqrt{5})$. This expands to

$1 - 2\sqrt{5} \times 2\sqrt{5} = 1 - 20 = -19$

$\dfrac{(1 - 2\sqrt{5}) \times (7)}{(1 - 2\sqrt{5}) \times (1 - 2\sqrt{5})}$ ← **Multiply 'top' and 'bottom' by $(1 - 2\sqrt{5})$**

$= \dfrac{7 - 14\sqrt{5}}{-19} = \dfrac{-7}{19} + \dfrac{14\sqrt{5}}{19}$ and this is an irrational number.

Remember
$(x + y)(x - y) = x^2 - y^2$.

PROGRESS
TEST

1 (a) Decrease 24 in the ratio 5 : 6

(b) Increase 3.5 in the ratio 7 : 5

(c) Decrease 210 in the ratio 2 : 3

2 Show whether these numbers are rational or irrational:

(a) $\sqrt{2} \times \sqrt{8}$ (b) $\dfrac{1}{4 - \sqrt{2}}$ (c) $\dfrac{\sqrt{20}}{\sqrt{5}}$ (d) $\dfrac{1 + \sqrt{2}}{1 - \sqrt{2}}$

2 (a) Rational, 4   (b) Irrational, $\dfrac{4 + \sqrt{2}}{14}$ (c) Rational, 2   (d) Irrational $-3 - 2\sqrt{2}$

1 (a) 20   (b) 4.9   (c) 140

1.9 *Calculator methods*

LEARNING
SUMMARY

After studying this section, you will be able to:

● *understand your calculator's keys and display*

● *understand exponential growth and decay*

● *find upper and lower bounds*

Understanding your calculator's keys and display

KEY POINT Calculators are not all the same. You need to be familiar with yours, so you may need to refer to your calculator manual.

You need to be able to do these tasks – test yourself:

● Key in 13 → store in **memory** → work out 16 ÷ 4 on the calculator → add the contents of the memory to the answer → your display should show 17.

● Use your calculator **bracket** (parenthesis) **keys** to work out ((4 + 1) × 4) ÷ 10 – your display should show 2.

● Use your calculator **power key** (sometimes labelled x^y or ^) to work out these approximations to π: $\left(\dfrac{77\,729}{254}\right)^{\frac{1}{5}}$ (π ≈ 3.141 592 65 ...) and $\sqrt{2} + \sqrt{3}$ (π ≈ 3.146 264 ...).

● Use the **exponent key** (sometimes labelled 'EXP') to give standard index form – to work out $(3.5 \times 10^3)(1.1 \times 10^2)$ $(= 3.85 \times 10^5)$.

● You will also need to be able to use **sin**, **cos**, **tan** and their inverses, and some statistical functions (such as **mean** and **standard deviation**).

Exponential growth and decay

KEY POINT

Exponential growth (or decay) occurs when a quantity grows by being multiplied by the same number, for example 2 4 8 16 32 64 ... (× **2** each time)

Or 100 10 1 0.1 0.01 0.001 ... (× **0.1** each time).

It also occurs when proportional or percentage changes are repeated.

For example, a young tree grows 20% in height each year. If the tree was 1 m tall when it was planted, how tall will it be after 10 years?

> Each year the height is **1.2** × (previous year's height).

After 1 year the tree is **1.2** × 1 m tall (for 20% increase the multiplier is 1.2).

After 2 years the tree is **1.2** × (**1.2** × 1) = $1 × 1.2^2$ m tall.

After 3 years the tree is **1.2** × [**1.2** × (**1.2** × 1)] = $1 × 1.2^3$ m tall.

> The power key was used to work this out.

So after 10 years the height will be $1 × 1.2^{10} = 6.19$ m (to 3 s.f.).

Loans such as mortgages charge **compound interest** – interest is paid on the amount borrowed **and** on the interest from the previous year. So £40 000 borrowed at 4% compound interest will (assuming no money is paid off the loan), after 5 years, be a debt of £40 000 × 1.04^5.

Upper and lower bounds

KEY POINT

Measurements are not exact.

When a number is stated to a certain accuracy, the greatest it could be is the upper bound and the least it could be is the lower bound.

For example, a length of 15 cm measured to the nearest cm could be anywhere in the range 14.5 cm to 15.499 999 999 999 9 ... or 14.5 ⩽ actual length < 15.5.

We say that the **upper bound** of the measured value is 15.5 and the **lower bound** is 14.5.

Examples

(a) Write down the upper and lower bounds of each of these values given to the accuracy stated:

  (i)   8 m (1 s.f.)   Lower bound 7.5 m, upper bound 8.5 m.

  (ii)   300 g (1 s.f.)   Lower bound 250 g, upper bound 350 g.

  (iii) 25 min (2 s.f.) Lower bound 24.5 min, upper bound 25.5 min.

(b) James has 40 identical marbles. Each weighs 65 g to the nearest gram.

  (i)   What is the upper bound for the weight of one marble?

         Answer: 65.5 g

  (ii)   What is the upper bound for the weight of the 40 marbles?

         Answer: $65.5 \times 40 = 2620$ g

(c) A model car travels 40 m, measured to one significant figure, at a speed of 2 m/s, measured to one significant figure. Between what limits does the time taken lie?

   The bounds for the distance are 35 m and 45 m and for the speed 1.5 m/s and 2.5 m/s.

   $$\text{Time taken} = \frac{\text{distance}}{\text{speed}},$$

   so the upper bound is $\frac{45}{1.5} = 30$ s and the lower bound is $\frac{35}{2.5} = 14$ s.

   $$\frac{\underline{\text{Distance upper bound}}}{\text{Speed lower bound}}$$

> **Ensure you divide and multiply in the right situations:**
> **speed = distance ÷ time,**
> **distance = speed × time and**
> **time = distance ÷ speed.**

PROGRESS TEST

1 Find the upper and lower bounds for $\frac{15}{6 \times 4}$.

   Each number is correct to the nearest whole number.

2 In 2000 the world population was 6 billion. This is growing by 1.2% every year. What will the world population be in 2050?

   Assume this constant factor increase every year for 50 years.

3 The population of a bacteria colony grows at a rate of 15% per hour.

   The population is measured as 2500 to start with.

   Calculate the bacteria population 4 hours after the first measurement.

1 Upper bound: 0.805 ... and lower bound: 0.495 ... so sensible answers would be 0.8 and 0.5.
2 10.9 billion (3 s.f.) or 11 billion to 2 s.f.
3 $2500 \times 1.15^4 = 4372.51$ ... – a sensible answer would be 4400.

1.10 Solving numerical problems

 LEARNING SUMMARY

After studying this section, you will be able to:

- *consider strategies for solving problems*
- *check your answers*

Strategies

All GCSE papers have some multi-step or unstructured questions. These may involve several steps and may not be so obvious to solve as some more direct questions.

Before starting questions like these, consider all the methods you know which may be relevant to the problem, and then select the most appropriate. Some students find that it helps to underline the key words, to make the question more transparent.

Checking

Always try to leave time to check your answers.

Always ask yourself – '*Does the answer seem reasonable?*' – for example, beds that are 15 m high or probabilities greater than 1 should set off warning bells.

Check that your answer fits the numbers in the question – for example if £400 is to be shared and the total of all the shares, including your answer, is greater than £400, you must have made a slip somewhere.

There will nearly always be marks for giving units and an answer to a suitable degree of accuracy – make sure you don't forget these.

It is particularly important to check answers that you have obtained by using your calculator.

 KEY POINT

Ask yourself these questions about your calculator answer:

- **Is it reasonable?**
- **Does an approximate calculation give a similar-sized answer?**
- **Do you get the same answer if you work it out again?**
- **Do you get back to the starting number if you use inverse operations and work back from the answer?**

PROGRESS TEST

1 There are 14 pounds in a stone.
   There are 2.2 pounds in a kilogram.
   A man weighs 12 stone 5 pounds.
   Calculate his weight in kilograms, to the nearest kilogram.

2 Two cars go around a race track. The first car takes 1 minute 10 seconds to complete a circuit and the other 1 minute 17 seconds.
   They start together on the starting line.
   Find how long it takes before they are together again.

3 A farm has an area of 234 hectares of land.
   $\frac{1}{3}$ of the land is pasture and $\frac{2}{9}$ is arable.

   The rest is woodland.
   Calculate the area of woodland.

4 A holiday costs £450 this year. The same holiday next year costs 20% more.
   What percentage discount should a company give so that customers who buy a holiday for next year now don't pay any extra?

5 The number 750 can be written as $2 \times 5^a \times b$, where a is a whole number and b is a prime number. What are the values of a and b?

6 In a horse race there was a very close finish. Amanda was a third of a length ahead of Bob and Bob was half a length ahead of Celina. By what fraction of a length did Amanda beat Celina?

1 79 kg
2 After $2 \times 5 \times 11 \times 7$ seconds = 12 min 50 s.
3 104 hectares
4 16.667% ≈ 17%
5 $a = 3$, $b = 3$
6 $\frac{5}{6}$ lengths

Number

Sample GCSE questions

1 Rumours spread very fast in a crowd.

A typical formula showing this is $n = 3^x$, where x is the time in minutes since the rumour began and n the number of people who heard it.

According to the formula, how many people will have heard the rumour five minutes after it began? **[3]**

$$Number\ of\ people = 3^5$$
$$= 3 \times 3 \times 3 \times 3 \times 3 \qquad ✔ ✔$$
$$= 243 \qquad ✔$$

Always show intermediate steps – there may be method marks available. The number of people increases exponentially.

2 The distance from Sheffield to Maidstone is 125 miles.

Amy drives from Sheffield to Maidstone in $2\frac{1}{2}$ hours.

Calculate her average speed for the journey. **[3]**

$$Average\ speed = \frac{distance}{time\ taken} = \frac{125}{2\frac{1}{2}} \qquad ✔$$

$$= 125 \div 2\frac{1}{2} \qquad ✔$$
$$= 50\ mph \qquad ✔$$

You need to know the speed, distance, time formula.

Check the answer – is 50 mph reasonable?

3 Work out

(a) $32.75 \div 0.25$ **[3]**

(b) The reciprocal of 0.4 **[1]**

(c) $3\frac{1}{3} - 1\frac{3}{4}$ **[3]**

(d) $3\frac{1}{3} + 1\frac{3}{4}$ **[3]**

(a) $32.75 \div 0.25 = 3275 \div 25$ ✔
$$= 131 \qquad ✔ ✔$$

Remove the decimal by multiplying both sides of the division by 100.

(b) $\frac{1}{0.4} = 2.5$ ✔

The reciprocal of n is $\frac{1}{n}$.

(c) $3\frac{1}{3} - 1\frac{3}{4} = \frac{10}{3} - \frac{7}{4}$ ✔

$$= \frac{40 - 21}{12} \qquad ✔$$

$$= \frac{19}{12} = 1\frac{7}{12} \qquad ✔$$

Adding and subtracting fractions is straightforward if you change the fractions into equivalent fractions with the same denominator.

(d) $3\frac{1}{3} + 1\frac{3}{4} = \frac{10}{3} + \frac{7}{4}$ ✔

$$= \frac{40 + 21}{12} \qquad ✔$$

$$= \frac{61}{12} = 5\frac{1}{12} \qquad ✔$$

When dealing with mixed numbers it is useful to convert them into improper or top heavy fractions (or deal with whole numbers separately).

Change improper factions into mixed numbers for the answer.

Sample GCSE questions

4 Find:

(a) The highest common factor of 48 and 72. **[2]**

(b) The least common multiple of 18 and 80. **[2]**

(a) $48 = 2^4 \times 3 \quad 72 = 2^3 \times 3^2$ ✔

$HCF = 2^3 \times 3 = 24$ ✔

(b) $18 = 2 \times 3^2 \quad 80 = 2^4 \times 5$ ✔

$LCM = 2^4 \times 3^2 \times 5 = 720$ ✔

> *When answering questions like these change each number into a product of its prime factors.*

5 Show that $(\sqrt{2} + \sqrt{8})^2$ is rational. **[3]**

$(\sqrt{2} + \sqrt{8})^2 = (\sqrt{2} + \sqrt{8})(\sqrt{2} + \sqrt{8})$ ✔

$= 2 + \sqrt{2}\sqrt{8} + \sqrt{2}\sqrt{8} + 8$

$= 2 + 2\sqrt{16} + 8$ ✔

$= 18 \text{ which is rational} \left(= \dfrac{18}{1}\right)$ ✔

> *With questions like these expand the brackets, then simplify.*

> *Remember $\sqrt{a} \times \sqrt{b} = \sqrt{ab}$.*

> *A rational number must be expressed as a fraction consisting of integers.*

6 Estimate the value of $\sqrt{\dfrac{40\,095}{9.87^2}}$.

Show clearly how you arrived at your answer. **[3]**

$\sqrt{\dfrac{40\,095}{9.87^2}} \approx \sqrt{\dfrac{40\,000}{100}}$ ✔

$= \sqrt{400}$ ✔

$= 20$ ✔

> *Always show each step in your approximation – making clear what you are doing by using '=' and '≈'.*

7 A sports charity shares an award of £5000 among three sports clubs. The money is to be shared in the ratio $10:8:7$. How much does each club get? **[3]**

$10 + 8 + 7 = 25 \text{ shares}$

$1 \text{ share is worth } 5000 \div 25 = 200$ ✔

$So \text{ the amounts are } £2000, £1600 \text{ and } £1400$ ✔✔

> *Total the shares, and then calculate what a single share is worth. Finally convert shares into money.*

8 (a) Given that $n = 0.\dot{1}2\dot{6}$, show that $999n = 126$. **[2]**

(b) Hence express $0.\dot{1}2\dot{6}$ as a fraction in its lowest terms. **[2]**

(a) $1000n = 126.126\,126\,126\,126\ldots$ ✔

$1000n - n = 999n = 126$ ✔

(b) $999n = 126$

$n = \dfrac{126}{999}$ ✔

$= \dfrac{14}{111}$ ✔

> *This is a standard method.*

> *Don't forget the question asked for a fraction in its lowest terms.*

Sample GCSE questions

9 The population of the UK is 6×10^7 people.

The UK has an area of 2.4×10^5 km^2.

What is the average number of people per km^2.

Give your answer as a 'normal' integer. **[3]**

$$Average\ number\ of\ people\ per\ km^2 = \frac{6 \times 10^7}{2.4 \times 10^5}$$ ✔

$$= 2.5 \times 10^2$$ ✔

$$= 250$$ ✔

> Show every step in your calculation – just in case.

10 **(a)** The expression $\sqrt{8} (1 + \frac{1}{3^2} + \frac{1}{5^2})^{0.5}$ gives an approximate value for π.

Find this value correct to 3 s.f. **[3]**

(b) Another approximation for π is $\sqrt{10}$.

What is the percentage difference, to 2 d.p., between this value and the accurate value of π given by your calculator? **[3]**

> Never just perform the calculation on the calculator and merely write down the final answer – write down the intermediate answers whenever you can.

(a) $\sqrt{8} (1 + \frac{1}{3^2} + \frac{1}{5^2})^{0.5} = \sqrt{8} (1.151\ 111\ 1...)$ ✔

$$= 3.034\ 615\ .....$$ ✔

$$= 3.03\ to\ 3\ s.f.$$ ✔

(b) $\%\ difference = \frac{(\pi - \sqrt{10})}{\pi} \times 100$

$or\ \frac{(\sqrt{10} - \pi)}{\pi} \times 100$ ✔

$$= \frac{(3.162\ 277... - 3.141\ 592...)}{3.141\ 592...} \times 100$$ ✔

> The percentage difference is the % of the calculator value of π.

$$= 0.66\%\ to\ 2\ d.p.$$ ✔

11 The population of the world in 2005 was 6.5 billion.

The annual population growth rate for the first half of the 21st century is expected to be 0.8%.

Given this growth rate, calculate the world population in 2025.

Give your answer to a sensible accuracy. **[4]**

6.5×1.008^{25} ✔✔

$= 793\ 280...$ ✔

$giving\ 8\ billion\ as\ a\ sensible\ answer.$ ✔

> The multiplier is 1.008.

> With the uncertainty in the world population, given as 6.5 billion – one or perhaps two significant figures in the answer is most appropriate.

Sample GCSE questions

12 (a) UK £1 coins are made from copper, zinc and nickel in the ratio by mass of 140 : 49 : 11.

(i) Calculate the percentage nickel in a £1 coin. **[2]**

(ii) £1 coins have a mass of 9.5 g.
What mass of a £1 coin is copper? **[2]**

(b) The Royal Mint states the mass of 5p coins to be 3.25 g.
What is the lower bound in mass for £1 worth of 5p coins?
State any assumption you make. **[3]**

(a) (i) $\% \ nickel = \dfrac{11}{140 + 49 + 11}$ ✔ ⟵ *11 out of 200 parts of the £1 coin are nickel.*

$= 5.5$ ✔

(ii) $\dfrac{140}{200} \times 9.5$ ✔ ⟵ *140 parts out of 200 of the 9.5g coin are copper.*

$= 6.65 \ g$ ✔

(b) *The mass of the 5p coin is given to 3 s.f. as 3.25 assuming an uncertainty of 0.5 in the least significant digit.* ✔

The lower bound of the mass of a 5p coin is 3.245 ✔ ⟵ *It is usual to assume an uncertainty of half a unit in the least significant digit – but you need to state this.*

So the lower bound for 20 will be 20 × 3.245 = 64.9 g ✔

Exam practice questions

1   (a)   Write 72 as a product of prime factors.   [3]
    (b)   Write down the LCM of 72 and 48.   [2]
    (c)   Simplify the following.

    (i) $16^{-\frac{1}{2}} \times 27^{\frac{2}{3}}$ (ii) $(\sqrt[3]{8})^2$ [4]

2   Calculate the following, giving your answers as fractions in their lowest terms.
    (a) $5\frac{1}{2} + 4\frac{1}{3} + 3\frac{1}{5}$ [3]
    (b) $7\frac{2}{5} - 6\frac{5}{8}$ [3]
    (c) $1\frac{2}{3} \div 2\frac{2}{9}$ [3]

3   The mass of a carbon atom is 2×10^{-23} g.
    How many carbon atoms are there in 10 g of pure carbon?   [3]

4   Work out $(2 - \sqrt{3})^2$, giving your answer in the form $a + b\sqrt{3}$ and state the values of
    a and b.   [2]

5   A sponge ball is dropped from a height of 250 cm.
    On each bounce the ball rises to 20% of its previous height.
    To what height will it rise on the third bounce?   [4]

6   Amy is going on holiday. She weighs her case on her bathroom scales, which
    weigh to the nearest kilogram. She finds her case weighs 18 kg.
    (a)   What are the upper and lower bounds for the weight of her case?   [2]
    (b)   On the way to the airport she removes her portable DVD player from her case.
          At the airport the weight of her case is given as 17.6 kg to the nearest 0.1 kg.
          What is the heaviest her DVD player could be?   [2]

7   The height above the earth's surface, in metres, at which a communication satellite must
    orbit in order to stay in position is given by the expression:

$$\sqrt[3]{\frac{(24 \times 3600)^2 \times 9.8 \times (6.4 \times 10^6)^2}{4\pi^2}} - 6.4 \times 10^3$$

    Evaluate the expression correct to 2 s.f.
    Give your answer in km.   [4]

8   A laptop computer costs £850 including VAT at 17.5%.
    What was its price before VAT was added?   [3]

2 Algebra

Overview

Topic	Section	Studied in class	Revised	Practice questions
2.1 Use of symbols	**Letter symbols**			
	Manipulation			
	Factorising			
2.2 Indices in algebra	**Using the rules of indices**			
2.3 Linear equations	**Solving linear equations**			
	Setting up linear equations			
2.4 Formulae	**Substituting into formulae**			
	Changing the subject of formulae			
2.5 Simultaneous linear equations	**Solving simultaneous linear equations using algebra**			
	Solving simultaneous linear equations using straight line graphs			
2.6 Quadratic equations	**Solving quadratic equations by factorising**			
	Solving quadratic equations by completing the square			
	Solving quadratic equations by using the formula			
	Solving simultaneous linear and quadratic equations			
2.7 Numerical and graphical methods of solving equations	**Trial and improvement**			
	Graphical methods to solve equations			
2.8 Inequalities	**Inequalities in one variable**			
	Inequalities in two variables			
2.9 Graphs of functions	**Equations of straight lines**			
	Graphs of non-linear functions			
2.10 Sequences	**Generating terms of a sequence**			
	Finding the nth term			
2.11 Direct and inverse proportion	**Solving problems involving proportion**			
	Graphs of proportional relationships			
2.12 Interpreting graphical information	**Graphs of real-life situations**			
2.13 Circles and equations	**Equation of the circle**			
2.14 Transformation of functions and their graphs	**Functions**			
	Transforming graphs of functions			
2.15 Algebraic proof	**Using algebra in proof**			

2.1 Use of symbols

LEARNING SUMMARY

After studying this section, you will be able to:
- *understand the different uses of letters as symbols*
- *understand some of the vocabulary used in algebra*
- *manipulate algebraic expressions*
- *factorise algebraic expressions*

Letter symbols

KEY POINT

Letters are used to stand for or represent:

- Unknown number(s) in equation(s), for example:

  $2x - 6 = 17$ (x is the unknown)

  $x - y = 7$ and $2x + y = 8$ (x and y are the unknowns)

- **Variables** in formulae, which can take many values, for example:

  $s = vt$

  $v^2 = u^2 + 2fs$

- Numbers in an **identity**, which can take any value, for example:

  $2(x + 1) \equiv 2x + 2$

  which is true for any value of x.

The symbol $\equiv$ means 'identical to': $2(x + 1) \equiv 2x + 2$ is called an **identity**.

Conventions

x is never written as $1x$.

Order of multiplication does not alter the answer:

$a \times 2 = 2 \times a = 2a$ – by convention we write 'number' followed by the letters in alphabetical order.

For example, $2 \times b \times a \times c$ is written as $2abc$.

Several words have a special meaning in algebra:

- **Term**: means a single group of symbols. Examples are: xy, $2a$, $-x^2$ and $3xy^2$. Like terms can be combined, so that $3a + 5ab + 4a - 2ab$ can be simplified by combining like terms to give $7a + 3ab$.
- **Expression**: any arrangement of letter symbols and possibly numbers. Examples are: $2a + 6b + 8$, $x^2 - 4x$, $a + 2b + 3c$.
- **Function**: a relationship between two sets of values such that a value from the first set maps on to a unique value in the second set. Examples are: for the function $y = x^2$, when $x = 5$, then $y = 5^2$ or 25; $y = 2x + 4$ is a function of x.

Manipulation

Remember $a \times a = a^2$, $b \times b \times b = b^3$, $ab^2 \times ab = a^2b^3$ and so on.

The manipulation of symbols obeys similar rules to those used in arithmetic:

$7a + 3a = 10a$ $8b - b = 7b$ $c - 3c = -2c$ $4(a + b) = 4a + 4b$ $10e \div 2 = 5e$

$4b \times 5a$ can be written out as $4 \times b \times 5 \times a$.

Order of multiplication does not matter, so this expression can be written as $4 \times 5 \times a \times b$, which simplifies to $20ab$.

When you divide the top and bottom of a fraction by the same number the result stays the same – division by the same number is sometimes called **cancelling down**.

A division can always be written as a fraction:

$3xy \div 12y = \dfrac{3xy}{12y} = \dfrac{x}{4}$ and $3x^3 \div x^2 = \dfrac{3x^3}{x^2} = 3x$

Brackets can be used to show that the operation outside the bracket applies to all the terms inside the brackets. For example, $a(a + b)$ means multiply $a + b$ by a, so $a(a + b) = a^2 + ab$

Brackets can be multiplied by other brackets.
Every term in one bracket must be **multiplied by every term** in the other bracket and then like terms collected.

Brackets can help with your algebra: $\dfrac{y}{4} = x + 3$ can be written as $\dfrac{y}{4} = (x + 3)$.

This makes 'seeing' $y = 4(x + 3)$ easier.

More marks are lost in algebra questions by slips when dealing with negative terms than anything else!

Examples

Expand the brackets, then collect like terms.

(a) $(a + b)(c + d) = a(c + d) + b(c + d) = ac + ad + bc + bd$

(b) $(x - 5)(x + 2) = x(x + 2) - 5(x + 2)$
$= x^2 + 2x - 5x - 10 = x^2 - 3x - 10$

(c) $(x - 5)^2 = (x - 5)(x - 5)$
$= x(x - 5) - 5(x - 5)$
$= x^2 - 5x - 5x + 25 = x^2 - 10x + 25$

PROGRESS TEST

1 Simplify these expressions, where possible:

(a) $5ab + 2a + b + 6b$ (b) $3x - 2y - 4x + y$ (c) $mn \times no$

(d) $3a \times 4ab$ (e) $\frac{1}{2}bc^2 \times 6ab$ (f) $\dfrac{6ab}{2b^2}$

(g) $n^2 + 2n$ (h) $2x^2 - x^2y + xy^2$

(i) $2x^2 + 3xy - 4y^2 + 2xy - 3x^2 + y^2$

2 Remove the brackets from these expressions and simplify where possible:

(a) $4a(3a - 6)$ (b) $(s + 4)(t - 2)$ (c) $(x - 4)(x + 3)$

(d) $(2x - 7)(5x - 3)$ (e) $(x + 1)^2$ (f) $(x - 1)^2$

1 (a) $5ab + 2a + 7b$ (b) $-x - y$ (c) mn^2o (d) $12a^2b$

(e) $3ab^2c^2$ (f) $\dfrac{3a}{b}$ (g) $n^2 + 2n$ (or $n(n + 2)$ or $x(2x - xy + y^2)$)

(h) $2x^2 - x^2y + xy^2$ (i) $-x^2 + 5xy - 3y^2$

2 (a) $12a^2 - 24a$ (b) $st + 4t - 2s - 8$ (c) $x^2 - x - 12$ (d) $10x^2 - 41x + 21$

(e) $x^2 + 2x + 1$ (f) $x^2 - 2x + 1$

Factorising

When dealing with algebraic expressions it is sometimes useful to 'think with numbers'.

$$10 \ (= 2 \times 5) \text{ and } 6 \ (= 2 \times 3) \text{ both have 2 as a common factor, so}$$

$$10 + 6 = 2(5 + 3)$$

In other words $10 + 6$ can be factorised to $2(5 + 3)$.

Check any factorisation by multiplying out (expanding) the factored expression – you should get back to the original expression.

You know that $x(a + b) = ax + bx$.

The reverse of this process, going from $ax + bx$ to $x(a + b)$, is called **factorising**. The terms $ax + bx$ have a common factor x.

Examples

(a) Factorise the expression $6n^2 + 2n$.

Each term of the expression can be divided by $2n$ (we say that $2n$ is a factor of each term).

So the answer is $2n(3n + 1)$.

(b) Factorise $3x + 2ax + 3y + 2ay$.

$$3x + 3y + 2ax + 2ay$$

| Group together similar terms |

$$3(x + y) + 2a(x + y)$$

| Factorise each pair of terms |

$$(3 + 2a)(x + y)$$

| $(x + y)$ is a common factor |

(c) Factorise $x^2 + 5x + 6$.

The answer is $(x + 2)(x + 3)$, which is best found by being methodical, rather than by trial and error.

If $x^2 + px + q$ factorises into $(x + a)(x + b)$ then by definition:

$(x + a)(x + b) = x^2 + px + q$

but

$(x + a)(x + b) = x^2 + xb + ax + ab = x^2 + (a + b)x + ab$

therefore $(a + b) = p$ and $ab = q$

So to factorise $x^2 + px + q$, you need to find two numbers that will add to give p and multiply to give q. Nothing has been said about whether a, b, p or q are negative or positive so we can assume that they can be either.

This method always works – if it is possible to factorise the expression.

Examples

(a) $x^2 + 5x + 6$

You need two numbers that add to give 5 and multiply to give 6.
Adding to give 5 would be (1 and 4) and (2 and 3) but only 2 and 3 multiply to give 6.
So
$x^2 + 5x + 6 = (x + 2)(x + 3)$

The main area for making errors is in dealing with negative numbers.

(b) $x^2 - 3x - 10$

You need two numbers that add to give −3 and multiply to give −10.
Adding to give −3 would be (0, −3), (1, −4), (2, −5), (3, −6) and so on.
Look at these to find a pair that multiply to give −10:
$0 \times -3 = 0$, $1 \times -4 = -4$, $\mathbf{2 \times -5 = -10}$
So $x^2 - 3x - 10 = (x + 2)(x - 5)$

Always quickly check by expanding your answer.

These are some common mistakes made when factorising:
- $16x^2 - 1$ is *not* equal to $(16x - 1)(16x + 1)$ – it *should be* $(4x - 1)(4x + 1)$.
- $x^2 + y^2$ cannot be factorised into the difference of two squares – it cannot be factorised at all.
- $6xy + 3xz = x(6y + 3z)$ can be factorised further into $3x(2y + z)$ – make sure you factorise fully.

You need to be able to recognise expressions that are differences of squares.

Expressions such as $x^2 - 100$, $x^2 - 4$ are called differences of squares. They factorise very easily:
$x^2 - 100 = (x - 10)(x + 10)$
$x^2 - 4 = (x - 2)(x + 2)$

Factorise these expressions:
1 $5c + 20d$
2 $4a^2 - 12a$
3 $4mn - 8n^2$
4 $ax + 2a + 2bx + 4b$
5 $x^2 + 7x + 12$
6 $x^2 - 8x + 16$
7 $x^2 - x - 20$
8 $x^2 - 49$

1 $5(c + 4d)$ **2** $4a(a - 3)$ **3** $4n(m - 2n)$ **4** $(x + 2)(a + 2b)$ **5** $(x + 3)(x + 4)$ **6** $(x - 4)^2$ **7** $(x - 5)(x + 4)$ **8** $(x - 7)(x + 7)$

 **Algebra**

2.2 Indices in algebra

 LEARNING SUMMARY

After studying this section, you will be able to:
- **divide and multiply algebraic expressions using indices**
- **find roots using indices**

Using the rules of indices

KEY POINT

These are the basic laws of indices:

$$x^m \times x^n = x^{m+n}$$
$$x^m \div x^n = x^{m-n}$$
$$(x^m)^n = x^{mn}$$
$$x^0 = 1$$
$$\frac{1}{x^n} = x^{-n}$$
$$\sqrt[n]{x} = x^{\frac{1}{n}}$$

Examples

(a) $\sqrt{\dfrac{x^6}{y^{12}}} = \left(\dfrac{x^6}{y^{12}}\right)^{\frac{1}{3}} = \dfrac{x^2}{y^4}$

(b) $(a^4 b^{10}) \div (ab^2) = a^{4-1} \times b^{10-2} = a^3 b^8$

(c) $(xy^3)^4 = x^4 y^{12}$

(d) If $x^2 = a^2 - 16$ then $x = \sqrt{a^2 - 16}$ not $a - 4$: always be careful when square rooting expressions. When in doubt, try with simple numbers, for example $\sqrt{3^2 + 4^2} = \sqrt{9 + 16} = \sqrt{25} = 5$ (not $3 + 4 = 7$).

 PROGRESS TEST

1 Simplify:
  (a) $\sqrt{a^6 b^6}$ (b) $(x^3 y^6)^0$ (c) $(a^{\frac{1}{4}} b^2) \times (a^{\frac{1}{4}} b^3)$
2 Simplify:
  (a) $(x^{-2} y^2) \div (x^{-3} y^4)$ (b) $\dfrac{\sqrt{p^4 q^{-2}}}{p^2 q^{-4}}$ (c) $\sqrt[3]{a^3 b^9}$ (d) $\sqrt{\dfrac{a^6 b^5}{a^7 b^4}}$ (e) $\left(\sqrt{x^6 y^2}\right)^4$

2 (a) xy^{-2} (b) q^3 (c) ab^3 (d) $\sqrt{\dfrac{b}{a}}$ (e) $x^{12} y^4$

1 (a) $a^3 b^3$ (b) 1 (c) $a^{\frac{1}{2}} b^5$

2.3 Linear equations

LEARNING SUMMARY

After studying this section, you will be able to:
- solve linear equations
- set up a linear equation, using symbols, from given information

Solving linear equations

 KEY POINT

An equation contains two expressions that are equal. To keep this equality, both expressions must be treated the same:

- The same number is added to (or subtracted from) each side.
- Each side is multiplied (or divided) by the same number.

A **linear equation** is one in which there are no powers of the unknown greater than 1: $3x - 6 = 45$ and $3x + 3 = 2x + 8$ are linear equations, but $3x^2 - 6 = 45$ and $3x^2 + 3 = 2x + 8$ are *not* linear.

Examples

(a) $$6x = 12$$
$$x = 2$$ ← Divide each side by 6

(b) $$\frac{x}{5} = 3$$
$$x = 15$$ ← Multiply each side by 5

(c) $3(2a + 7) = 2(a - 1) + 19$

$$6a + 21 = 2a - 2 + 19$$ ← Expand the brackets

$$6a + 21 = 2a + 17$$

$$4a + 21 = 17$$ ← Subtract $2a$ from both sides

$$4a = -4$$ ← Subtract 21 from both sides

$$a = -1$$ ← Divide both sides by 4

When $a = -1$

LHS $\qquad 3(2a + 7) = 15$

RHS $2(a - 1) + 19 = -4 + 19 = 15$

LHS = RHS, correct solution.

> Check your solution by substituting it back into the original equation to ensure that both sides (expressions) are in fact equal.

(d) $4d + 2 = 2d - 7$

$$2d + 2 = -7$$ ← Subtract $2d$ from each side

$$2d = -9$$ ← Subtract 2 from each side

$$d = -4.5$$ ← Divide by 2 (don't forget the negative sign)

PROGRESS
TEST

Solve these equations. Check your answers by substituting your solution into the original equation.

1 $3(q + 1) = 2q$
2 $\frac{1}{3}(2x + 1) = \frac{1}{2}(x - 3)$
3 $\frac{1}{2}(3x - 5) = x + 10$
4 $4(5s + 1) = 14s - 1$

1 $q = -3$
2 $x = -11$
3 $x = 25$
4 $s = -\frac{5}{6}$

Setting up linear equations

From information given in words or on a diagram you will be asked to set up and solve an equation for an unknown quantity.

> Always state what your letter represents.

> Always check that your equation fits the given 'story' by re-reading it against your equation.

> When forming an equation, always make it clear what your letters represent and remember to check your answer by putting it into the story – does it fit?

Examples

(a) I think of a number, double it and add 8. The result is 50.
What number did I think of?
Let x represent the number, following the instructions above we get:

$2x + 8 = 50$
giving $x = 21$

(b) A bag contains white, blue and red counters. There 14 more blue counters than white and 6 fewer red counters than white. There are 44 counters altogether. How many white counters are there?
Let w represent the number of white counters.
There are $w + 14$ blue counters.
There are $w - 6$ red counters.
The total number of counters is 44.

So $w + w + 14 + w - 6 = 44$
$3w + 8 = 44$
$w = 12$, so there are 12 white counters.

 PROGRESS TEST

1 The length of a rectangular field is 10 metres more than its width. The perimeter of the field is 220 metres. What is the width of the field?
2 Jordan has only 5p and 50p coins in her money box. She has twice as many 50p coins as 5p coins. The total amount of money in her money box is £10.50. How many (a) 50p and (b) 5p coins are there in her money box?
3 Jake is x years old. His sister Simone is 4 years older. Their mother is twice as old as the total of their ages. The sum of all three people's ages is 72 years. How old is each of the three people?
4 There are two numbers. The first number is four more than twice the second number. The sum of the two numbers is 97. Find the two numbers.

4 31 and 66
3 10, 14 and 48 years old
2 (a) 20 50p coins   (b) 10 5p coins
1 50 m

2.4 Formulae

LEARNING SUMMARY

After studying this section, you will be able to:
- *substitute numbers into formulae*
- *change the subject of a formula*
- *construct formulae*

Substituting into formulae

> **KEY POINT**
>
> When substituting into formulae, check to see if you need to give the units of your answer. For these examples you have not been given any units, so there is no need.

Replace the letters in the formula with the given numbers then do the arithmetic.

Examples

(a) Use the formula $A = \dfrac{h(a+b)}{2}$ to find the value of A when $h = 10$, $a = 4$ and $b = 12$:

$$A = \frac{10(4+12)}{2} = 5 \times 16 = 80$$

(b) Find the value of v using the formula $v^2 = u^2 + 2fs$, when $u = 12$, $f = 5$ and $s = 12.5$:

$$v^2 = 12^2 + 2 \times 5 \times 12.5$$
$$= 144 + 125$$
$$= 269$$
$$\text{so } v = \sqrt{269} = 16.40 \text{ (to 2 d.p.)}$$

Changing the subject of formulae

>
>
> **KEY POINT**
>
> The rules for manipulating formulae are the same as for equations, including this one which you may not have met or used before: the equality remains true if you take a root or power of both sides.
>
> For example: $a^2 b^4 = 16t$.
>
> Taking the square root of each side gives $ab^2 = \pm 4\sqrt{t}$.
>
> In the case of $\sqrt{a^2 + b^2} = z$ squaring each side gives $a^2 + b^2 = z^2$.

Be systematic – work through one line at a time. Do not try to do two steps at once – remember method marks may be available.

Examples

(a) Make u the subject of the formula:

$$s = ut + \tfrac{1}{2}gt^2$$

$$ut + \tfrac{1}{2}gt^2 = s$$ ⟵ Swap sides to get u (the new subject) on to the left-hand side (if $x = ab$, then $ab = x$)

$$ut = s - \tfrac{1}{2}gt^2$$ ⟵ Subtract $\tfrac{1}{2}gt^2$ from each side

$$u = \frac{s - \tfrac{1}{2}gt^2}{t}$$ ⟵ Divide each side by t – this is where mistakes can be made

$$u = \frac{2s - gt^2}{2t}$$ ⟵ Remove the $\tfrac{1}{2}$ by multiplying top and bottom by 2

> If the new subject is involved in a bracket, expand the bracket as soon as possible.

(b) Make x the subject of the formula:

$$x + y = c(x + z)$$

$$x + y = cx + cz$$

$$x - cx + y = cz$$ ⟵ Collect on the left-hand side all the terms involving the new subject – by subtracting cx from each side

$$x - cx = cz - y$$ ⟵ Subtract y from each side and group the x terms by factorising the left-hand side

$$x(1 - c) = cz - y$$

$$x = \frac{cz - y}{1 - c}$$ ⟵ Divide both sides by $(1 - c)$

(c) Make s the subject of $t = \sqrt{\dfrac{3}{s + 2}}$:

$$t^2 = \frac{3}{s + 2}$$ ⟵ Square both sides

$$t^2(s + 2) = 3$$ ⟵ Multiply both sides by $(s + 2)$ to remove the fraction

$$st^2 + 2t^2 = 3$$

$$st^2 = 3 - 2t^2$$

$$s = \frac{3 - 2t^2}{t^2}$$

> Lay out your working so that the steps and the flow in your algebra are clear.

Always re-read the question to check that your formula matches the original information given. Sometimes it can be useful to substitute a few simple numbers into your formula and ask yourself, 'does the resulting answer make sense?'

You must know these two circle formulae.

Example

A circle has circumference C and radius r.

Find a formula for its area A in terms of C.

$$C = 2\pi r \qquad\qquad [1]$$
$$A = \pi r^2 \qquad\qquad [2]$$

From [1] $r = \dfrac{C}{2\pi}$

Substituting this value of r into [2] gives $A = \pi \left(\dfrac{C}{2\pi}\right)^2$

$$= \dfrac{\pi C^2}{4\pi^2} = \dfrac{C^2}{4\pi}$$

It's useful when using more than one formula to label them – it makes your working clearer.

Break up the working into simple steps.
Simple check: the units of the answer are [length]² which is the unit of area – which fits.

PROGRESS TEST

1 Substitute into these formulae:
  (a) Find s when $s = ut - \frac{1}{2}gt^2$ and $u = 1$, $g = 10$ and $t = 5$.

  (b) Find T when $T = 2\pi\sqrt{\dfrac{l}{g}}$ and $l = 1.5$, $g = 10$, π is taken as 3.1.

2 (a) Make l the subject of the formula $T = 2\pi\sqrt{\dfrac{l}{g}}$.

  (b) Make x the subject of the formula $x^2 + y^2 = r^2$.
  (c) Make n the subject of the formula $3n - m = n(m - 1)$.

  (d) Make u the subject of the formula $a = \dfrac{bu + v}{u - c}$.

  (e) Make a the subject of the formula $a^2 b^2 = a^2 + b^2$

3 (a) The volume, V, of a sphere, radius r, is given by $V = \frac{4}{3}\pi r^3$, and the surface area, S, by $S = 4\pi r^2$. Find the formula for V in terms of S.
  (b) These two formulae are used in mechanics:
  $E = \frac{1}{2}mv^2$ and $P = mv$
  Construct a formula for E in terms of p and m.

3 (a) $V = \dfrac{S^{1.5}}{6\sqrt{\pi}}$ (b) $E = \dfrac{p^2}{2m}$ ($v = \dfrac{p}{m} \rightarrow v^2 = \dfrac{p^2}{m^2}$ then put this in $E = \frac{1}{2}mv^2$...)

2 (a) $l = \dfrac{T^2 g}{4\pi^2}$ (b) $x = \sqrt{r^2 - y^2}$ (c) $n = \dfrac{4 - m}{m}$ (d) $u = \dfrac{ac + v}{a - b}$ (e) $a = \sqrt{\dfrac{b^2}{b^2 - 1}}$

1 (a) -120 (b) 2.4 to 1 d.p.

2.5 Simultaneous linear equations

After studying this section, you will be able to:

- *solve simultaneous linear equations by substitution*
- *solve simultaneous linear equations by elimination*
- *solve simultaneous linear equations using straight line graphs*

Solving simultaneous linear equations using algebra

If an equation has two unknowns, such as $2y + x = 20$, it cannot have unique solutions. Two unknowns require two equations which are solved at the same time (simultaneously) – but even then two equations involving two unknowns do not always give unique solutions.

With this method you need to take particular care with your algebra. Should your solutions be 'strange' fractions such as $\frac{9}{13}$ the chances are you've made a slip – check your algebra.

This method is called **solution by substitution**.

It involves what it says – substitution – using one of the equations to get an expression of the form '$y = \ldots$' or '$x = \ldots$' and substituting this into the other equation. This gives an equation with just one unknown, which can be solved in the usual way. This value is then substituted in one or other of the original equations, giving an equation with one unknown.

It is a good idea to label each equation. It helps you explain what you are doing – and may gain you method marks.

This value of x can be substituted into equation [1] or [2], or into the expression for y: $y = 2x - 1$. Choose the one that is easiest!

As a check, substitute the values back into each of the two starting equations.

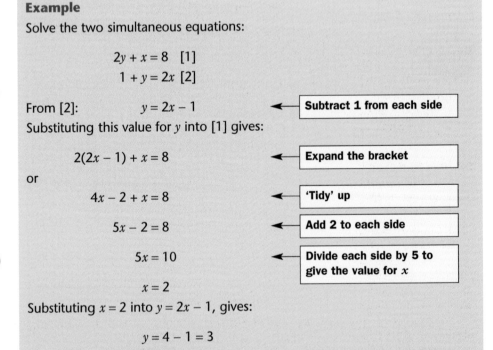

Example

Solve the two simultaneous equations:

$$2y + x = 8 \quad [1]$$
$$1 + y = 2x \quad [2]$$

From [2]: $\qquad y = 2x - 1$ ◄ **Subtract 1 from each side**

Substituting this value for y into [1] gives:

$$2(2x - 1) + x = 8$$ ◄ **Expand the bracket**

or

$$4x - 2 + x = 8$$ ◄ **'Tidy' up**

$$5x - 2 = 8$$ ◄ **Add 2 to each side**

$$5x = 10$$ ◄ **Divide each side by 5 to give the value for x**

$$x = 2$$

Substituting $x = 2$ into $y = 2x - 1$, gives:

$$y = 4 - 1 = 3$$
$$y = 3$$

So the two solutions are $x = 2$ and $y = 3$.

LHS [1] = 2(3) + (2) = 8   RHS [1] = 8   and   LHS [1] = RHS [1]

LHS [2] = 1 + (3) = 4   RHS [2] = 2(2) = 4   and   LHS [2] = RHS [2]

The second method is called **solution by elimination**.
It works because of two properties of equations:

- Multiplying (or dividing) the expression on each side by the same number does not alter the equation.
- Adding two equations produces another valid equation:
  e.g. $2x = x + 10$ ($x = 10$) and $x - 3 = 7$ (x also $= 10$).
  Adding the equations gives $2x + x - 3 = x + 10 + 7$ (x also $= 10$).

The object is to manipulate the two equations so that, when combined, either the x term or the y term is eliminated (hence the name) – the resulting equation with just one unknown can then be solved:

> **The method is not quite as hard as it first seems, but it helps if you know why it works.**

> **We want to manipulate one of the equations – [1] – so that when it is combined (either added or subtracted) with the other, either the xs or the ys drop out. In this example, the x terms dropped out, giving a solution for y. To find the value of x, the value found for y is substituted into one of the original equations.**

$$2y + x = 8 \qquad [1]$$

$$1 + y = 2x \qquad [2]$$

$$y - 2x = -1 \qquad [2] \text{ rearranged}$$

$[1] \times 2$ gives $\qquad 4y + 2x = 16 \qquad [3]$
so $[2] + [3]$ gives $\qquad 5y = 15$
$$y = 3$$

> **Label your equations.**

> **Rearrange one of the equations so that it is in a similar form to the other.**

Substituting this, $y = 3$ into [2] gives $1 + (3) = 2x$
so $2x = 4$, giving $x = 2$
So the two solutions are $x = 2$ and $y = 3$.

The solutions can be checked as in the previous example.

Solving simultaneous linear equations using straight line graphs

> **Because the graphs of $2y + x = 8$ and $y + 1 = 2x$ are straight lines, they are called linear equations.**

> **See page 71 if you are not sure how to plot straight line graphs.**

> **Drawing two straight line graphs to solve simultaneous equations also works the other way round – solving the two simultaneous equations algebraically gives the coordinates of the point of intersection of the two straight line graphs of the two simultaneous equations.**

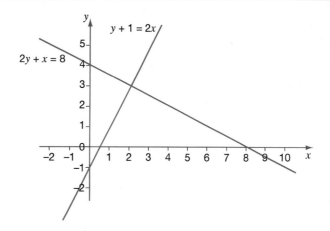

The two straight lines represent the equations $2y + x = 8$ and $y + 1 = 2x$.
This means, for example, that for any point on the line $2y + x = 8$, 2 times the value of y plus the value of x gives 8; similarly for $y + 1 = 2x$.

There is only one point that fits both equations. This is the point (2, 3) where the two straight lines cut.

So $x = 2$ and $y = 3$ are the solutions to the simultaneous equations $2y + x = 8$ and $y + 1 = 2x$.

Algebra

PROGRESS TEST

1 (a) Solve this pair of simultaneous equations by elimination:
$2x + y = 7$, $4x + 3y = 15$
Check your answers.

(b) Solve this pair of simultaneous equations by substitution:
$5p - 2q = 5$, $q = 2p - 1$
Check your answers.

2 Solve each of these pairs of simultaneous equations by a method of your choice:
(a) $3x + 4y = 10$, $x = 3 - y$
(b) $p - 2q = 4$, $3p + 2q = 8$
(c) $2y + x = 10$, $3x = -y$
(d) $a + b = -1$, $2a - b = 4$

3 Find, without drawing a graph, the coordinates of the point where the lines $5x + 2y = 16$ and $3x = y + 3$ meet.

4 Draw graphs to solve these equations:
(a) $x + y = 4$, $y = x - 2$
(b) $x + 2y = 4$, $y = x - 1$

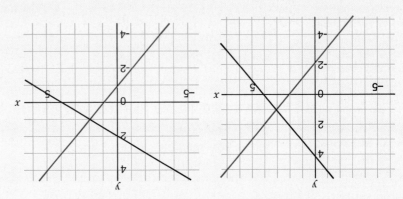

1 (a) $x = 3$, $y = 1$ (b) $p = 3$, $q = 5$
2 (a) $x = 2$, $y = 1$ (b) $p = 3$, $q = -0.5$ (c) $x = -2$, $y = 6$ (d) $a = 1$, $b = -2$
3 (2, 3) (solve the simultaneous equations $5x + 2y = 16$ and $3x = y + 3$)
4 (a) $x = 3$, $y = 1$.   (b) $x = 2$, $y = 1$.

2.6 Quadratic equations

LEARNING SUMMARY

After studying this section, you will be able to:
- *solve quadratic equations by factorising*
- *solve quadratic equations by completing the square*
- *solve quadratic equations by using the formula*
- *solve simultaneous equations when one of them is quadratic*

Solving quadratic equations by factorising

Unless a graphical method is asked for, quadratic equations on the non-calculator paper will probably involve factorising or completion of the square. Quadratic equations can have two different solutions or **roots**.

You may need a quick look at pages 50–51 to remind yourself how to factorise expressions such as:

$x^2 - x - 6$

which factorises into $(x - 3)(x + 2)$,

$a^2 - 3a$

which factorises into $a(a - 3)$

and

$b^2 - 2b + 1$

which will factorise into $(b - 1)^2$.

KEY POINT

If two numbers or expressions are multiplied and the result is zero, then one or both of them must be zero.

If $A \times B = 0$ then $A = 0$ or $B = 0$.

If $(x - 3)(x + 2) = 0$ then $(x - 3) = 0$ or $(x + 2) = 0$

and if $(x - 3) = 0$ then $x = 3$

or if $(x + 2) = 0$ then $x = -2$

Examples

(a) Solve $x^2 - x - 6 = 0$

Factorising gives

$(x - 3)(x + 2) = 0$

So either $(x - 3) = 0$ in which case $x = 3$
or $(x + 2) = 0$ in which case $x = -2$.
This means that $x = 3$ and $x = -2$ are both solutions (roots) to
$x^2 - x - 6 = 0$.

(b) Solve $a^2 - 3a = 0$

Factorising gives

$a(a - 3) = 0$

So either $a = 0$ or $(a - 3) = 0$ in which case $a = 3$.
The two solutions are $a = 0$ and $a = 3$.

(c) Solve $b^2 - 2b + 1 = 0$

Factorising gives

$(b - 1)^2 = 0$

So $(b - 1) = 0$, in which case $b = 1$.
In this case there are two identical solutions $b = 1$.
The equation has **repeated roots**.

Solving quadratic equations by completing the square

Example

$$(x + 3)^2 - 4 = 0$$

$$(x + 3)^2 = 4$$ ← | Add 4 to each side |

$$(x + 3) = \pm 2$$ ← | Take the square root of each side, $\sqrt{4} = \pm 2$ $(-2 \times 2 = 4, 2 \times 2 = 4)$ |

so $x = -3 \pm 2$

Check by substituting both roots back into the original equation.

$$x = -1 \text{ and } x = -5$$

KEY POINT

Completing the square involves rearranging a quadratic equation into the form

$$(x + a)^2 - b = 0$$

where a and b are numbers, so that

$$(x + a)^2 = b$$

Taking the square root of both sides gives

$$(x + a) = \pm\sqrt{b}$$

giving $x = \pm\sqrt{b} - a$ ($\pm\sqrt{b}$ gives two roots)

Completing the square can be used to give answers to a given accuracy or in **surd** form (e.g. $x = \sqrt{5}$, $x = \sqrt{3}$).

This is quite a common way to lead into asking you to use completion of the square.

Remember in, for example, $(x + n)^2$ the number of xs (called the **coefficient** of x) is $2n$. So the coefficient of x will be 6 in $(x + 3)^2$.

Example

Rewrite $x^2 + 6x - 15$ in the form $(x + p)^2 - q$.

Hence solve the equation $x^2 + 6x - 15 = 0$.
Leave your answer in surd form.

$(x + 3)^2$ will give an x^2 and $6x$ term

| Adjust $(x + 3)^2 = x^2 + 6x + 9$ to give $x^2 + 6x - 15$ | ←

$$(x + 3)^2 = x^2 + 6x + 9$$

$$x^2 + 6x - 15 = (x^2 + 6x + 9) - 24$$

So $x^2 + 6x - 15 = (x + 3)^2 - 24$ ← | The square has been completed |

giving $p = 3$ and $q = 25$

As a result of the above completion of the square, $x^2 + 6x - 15 = 0$ can be rewritten:

$$(x + 3)^2 - 24 = 0$$

or

$$(x + 3)^2 = 24$$

$$x + 3 = \pm\sqrt{24}$$ ← | Taking the square root of each side |

so

$$x = -3 \pm\sqrt{24} = -3 \pm\sqrt{4 \times 6} = -3 \pm\sqrt{4} \times \sqrt{6} = -3 \pm 2\sqrt{6}$$

Solving quadratic equations by using the formula

When using the quadratic formula, don't forget the '$2a$' denominator. Also, be careful when dealing with negative numbers inside the square root. State your values of a, b and c to be used in the formula.

KEY POINT

The solution to the equation $ax^2 + bx + c = 0$ is given by the equation

$$x = \frac{-b \pm \sqrt{b^2 - 4ac}}{2a}$$

If $(b^2 - 4ac)$ is negative there are no real solutions because this would mean taking the square root of a negative number.

If $b^2 = 4ac$ there is just one solution.

Easily checked – especially if your calculator can store numbers as variables.

Examples

(a) $x^2 - 3x - 13 = 0$ $a = 1$ $b = -3$ $c = -13$

$$x = \frac{-(-3) \pm \sqrt{(-3)^2 - 4(1 \times -13)}}{2 \times 1} = \frac{3 \pm \sqrt{9 + 52}}{2} = \frac{3 \pm \sqrt{61}}{2}$$

$x = -2.41$ and $x = 5.41$ (3 s.f.)

(b) $3x^2 - 9x + 5 = 0$ $a = 3$ $b = -9$ $c = 5$

$$x = \frac{-(-9) \pm \sqrt{(-9)^2 - 4(3 \times 5)}}{2 \times 3} = \frac{9 \pm \sqrt{81 - 60}}{6} = \frac{9 \pm \sqrt{21}}{6}$$

$x = 2.26$ and $x = 0.74$ (3 s.f.)

In the expression $4x^3 - x$ the **coefficient** of x^3 is 4, the coefficient of x^2 is 0, and the coefficient of x is -1.

A **root** is a solution to an equation, so -4 and 4 are both roots of the equation $x^2 = 16$.

A **surd** is an irrational number expressed as the root of a positive whole number. Examples are: $\sqrt{7}$, $4\sqrt{3}$, $3 + 4\sqrt{11}$, but $\sqrt{4} = 2$ is not a surd.

PROGRESS TEST

1 Solve these equations by factorising:
   (a) $x^2 - 7x + 12 = 0$ (b) $x^2 + 9x - 10 = 0$
   (c) $2x^2 + 6x - 80 = 0$ (d) $x^2 - 4x = 45$

2 (a) Write these expressions in the form $(x + a)^2 - b$:
      (i) $x^2 + 4x - 1$ (ii) $x^2 + 6x + 7$ (iii) $x^2 - 2x - 7$
   (b) Use your answers to (a) to solve these quadratic equations by completing the square. Leave your answers in surd form.
      (i) $x^2 + 4x - 1 = 0$ (ii) $x^2 + 6x + 7 = 0$ (iii) $x^2 - 2x - 7 = 0$

3 Solve these equations using the formula. Give your answers to 2 d.p.
   (a) $x^2 - 3x - 2 = 0$ (b) $2x^2 - 3x = 1$

1 (a) $x = 3$ or $x = 4$ (b) $x = 1$ or $x = -10$
   (c) $x = -8$ or $x = 5$ (d) $x = -5$ or $x = 9$
2 (a) (i) $(x + 2)^2 - 5$ (ii) $(x + 3)^2 - 2$ (iii) $(x - 1)^2 - 8$
   (b) (i) $x = -2 \pm \sqrt{5}$ (ii) $x = -3 \pm \sqrt{2}$ (iii) $x = 1 \pm \sqrt{8} = 1 \pm 2\sqrt{2}$
3 (a) $x = 3.56$ or -0.56 (b) $x = 1.78$ or -0.28

Solving simultaneous linear and quadratic equations

> **KEY POINT**
>
> Substitution is the method used here to solve a pair of simultaneous equations when one of them is quadratic. It is important to decide which equation to use to substitute into the other equation – this depends very much on the pair of equations and the algebra involved.
>
> - As in the case of two linear simultaneous equations, the point or points of intersection of the curve (quadratic equation) and the straight line (linear equation) give the solution(s) to the corresponding simultaneous equations.
>
> - The equation of a circle centre (0, 0) and radius r is given by:
>
> $$x^2 + y^2 = r^2$$

Example

Find the points of intersection of the circle $x^2 + y^2 = 100$ and the straight line $x - y = 2$.

Using $x - y = 2$ gives $x = y + 2$.

Substituting this value for x into the equation for a circle gives:

$(y + 2)^2 + y^2 = 100$

Expanding the brackets:

$y^2 + 4y + 4 + y^2 = 100$

Collecting like terms:

> **Divide each term by 2 to give a coefficient of y^2 of 1.**

$2y^2 + 4y - 96 = 0$

$y^2 + 2y - 48 = 0$

> **We now need to solve this quadratic equation for y: factorising seems to suggest itself.**

$(y - 6)(y + 8) = 0$

so

$y = 6$ or $y = -8$

> **It seems reasonable to assume there may be two points of intersection.**

Using the straight line equation $x = y + 2$, when $y = 6$, $x = 8$ and when $y = -8$, $x = -6$.

So the points of intersection are (8, 6) and (−6, −8).

Check the answers by substituting these values into the original equations.

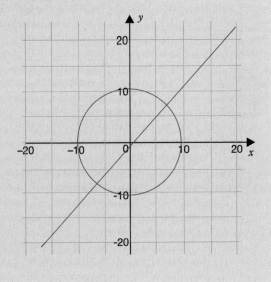

However, looking at the rough sketch above, the solution seems reasonable.

Example

Solve the simultaneous equations $y = x^2 - 2x + 3$, $y = 3x - 2$

Substituting $y = 3x - 2$ into $y = x^2 - 2x + 3$ gives

A brief inspection shows that this expression will not factorise so the formula or completing the square are the appropriate methods.

$3x - 2 = x^2 - 2x + 3$

or

$x^2 - 5x + 5 = 0$

Using the formula with $a = 1$, $b = -5$ and $c = 5$, $x = \dfrac{5 \pm \sqrt{5}}{2}$ so $x = 1.38$ or 3.62 (to 3 s.f.).

Substituting these values of x into $y = 3x - 2$ gives the corresponding values of y:

$y = 2.14$ or $y = 8.86$

So the points of intersection are $(1.38, 2.15)$ and $(3.62, 8.85)$.

Remember the formula will still work even if the LHS of the equation can be factorised – but in most cases it is easier to factorise.

PROGRESS TEST

1 Solve these pairs of simultaneous equations. Give your answers to 3 s.f.
   (a) $y = x^2 - 3$, $x = y - 9$ (b) $y = (x - 1)^2$, $y = x + 1$
   (c) $x^2 + y^2 = 16$, $y = x - 1$ (d) $y = x^2$, $x + y = 3$
2 Try to find the point(s) of intersection of the curve $y = x^2 + x + 2$ with the straight line $y = 2x - 1$. What does your result tell you about the curve and the straight line?

1 (a) $x = 4$, $y = 4$, or $x = -3$, $y = 6$ (b) $x = 0$, $y = 1$ or $x = 3$, $y = 4$
   (c) $x = -2.28$, $y = -3.28$ or $x = 3.28$, $y = 2.28$ (d) $x = -2.30$, $y = 5.30$ or $x = 1.30$, $y = 1.7$
2 Substituting $y = 2x - 1$ into $y = x^2 + x + 2$ results in the quadratic $x^2 - x + 3 = 0$ using the
   formula $x = \dfrac{1 \pm \sqrt{11}}{2}$ which is not allowed – so the curve and straight line cannot intersect.

2.7 Numerical and graphical methods of solving equations

LEARNING SUMMARY

After studying this section, you will be able to:
* *solve equations that cannot be solved with simple manipulative methods by using systematic trial and improvement in various forms*
* *solve equations that cannot be solved with simple manipulative methods by using graphical methods*

Trial and improvement

Sometimes it may not be possible to solve an equation using simple algebra, for example $x^3 + 2x^2 - 5 = 0$. However, there are approximate methods that can be refined to give solutions to any degree of accuracy. The two most important approaches involve **trial and improvement** and **graphical methods**.

 **Algebra**

> **KEY POINT**
> The object of the method of **trial and improvement** is to find systematically a value or values of x that makes the expression $x^3 + 2x^2 - 5 = 0$ as close as possible to 0. This (these) value(s) are the root(s) of the equation.

The trials are chosen for a reason, not haphazardly.

With a trial of:

$x = 1$ the value of $x^3 + 2x^2 - 5 = 0$ is $(1)^3 + 2(1)^2 - 5 = -2$ which is too small

$x = 2$ the value of $x^3 + 2x^2 - 5 = 0$ is $(2)^3 + 2(2)^2 - 5 = 11$ which is too large.

So we know the root must lie between $x = 1$ and $x = 2$.

An obvious, more accurate, trial is $x = 1.5$ (half way between $x = 1$ and $x = 2$).

With a trial of:

$x = 1.5$, the expression has the value $(1.5)^3 + 2(1.5)^2 - 5 = 2.875$ – too large

$x = 1.4$, the expression has the value $(1.4)^3 + 2(1.4)^2 - 5 = 1.664$– still too large

$x = 1.3$, the expression has the value $(1.3)^3 + 2(1.3)^2 - 5 = 0.577$ – still too large

$x = 1.2$, the expression has the value $(1.2)^3 + 2(1.2)^2 - 5 = -0.392$– too small.

The root lies between $x = 1.24$ and 1.25

$x = 1.25$, the expression has the value $(1.25)^3 + 2(1.25)^2 - 5 = 0.078\,125$

$x = 1.24$, the expression has the value $(1.26)^3 + 2(1.26)^2 - 5 = -0.01\,818$

By choosing $x = 1.245$ we can find whether $x = 1.24$ or $x = 1.25$ is closer

So far we know that the root is between 1.24 and 1.25:

$x = 1.245$, the expression has the value $(1.245)^3 + 2(1.245)^2 - 5 = 0.029\,831$.

Greater accuracy can be achieved by continuing the process. A sensible next value to try would be $x = 1.244$.

As $x = 1.245$ is just too large, we have found that the root of the equation between 1 and 2 is 1.24 to 2 d.p.

Make sure you show your trials and the results of them – just the answer on its own will not gain you full marks. Be systematic in your choice of trials.

Graphical methods to solve equations

In some cases the graph will be drawn for you, but you still need to be able to draw graphs like this.

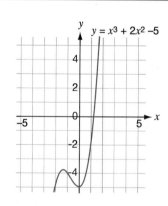

This is the graph of $y = x^3 + 2x^2 - 5$.

The roots of $x^3 + 2x^2 - 5 = 0$ are the values that make $x^3 + 2x^2 - 5 = 0$, which is the value of x that makes $y = 0$ on this graph – i.e. where it cuts the x-axis (or $y = 0$). This is about 1.2 (in line with the solution arrived at using trial and improvement).

On page 59 we found that the solution to two simultaneous equations was the coordinates of the point of intersection of the two equations drawn as straight lines or curves.

A common question is to give the graph of

$y = x^3 + 4x^2 - 5$

and to ask you to use it to solve, for example,

$x^3 + 4x^2 - x - 6 = 0$

The 'trick' is to re-write this equation so that one side is the given equation:

$x^3 + 4x^2 - 5 + (-x - 1) = 0$

which can be rearranged to give

$x^3 + 4x^2 - 5 = x + 1$

This can be solved using the given graph and plotting $y = x + 1$ as in this case.

The x-coordinates of the point of intersection of

$y = x^3 + 4x^2 - 5$ and $y = x + 1$ are the roots (three in this case) of the equation

$x^3 + 4x^2 - 5 = x + 1$

or

$x^3 + 4x^2 - x - 6 = 0$

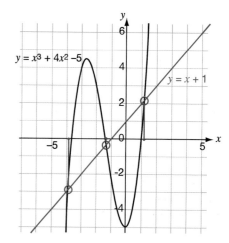

The approximate values to the three roots are −3.9, −1.2 and 1.1. More accurate values could be found by drawing the graph on an expanded scale – easy to do using a graphical calculator!

---

1 Use the method of trial and improvement to find a root for each of these equations, correct to two decimal places:
   (a) $x^3 - x - 2 = 0$ between 1 and 2
   (b) $x^3 + 5x = 50$, between 3 and 4
2 Use the method of trial and improvement to find a root for each of these equations, correct to one decimal place:
   (a) $x^4 + 3x - 6 = 0$
   (b) $3^x = 30$
3 (a) Draw the graph of the function $y = x^3 + 2x - 5$, for $-2 \leqslant x \leqslant 2$.
   (b) Use your graph to find approximate values for all the roots of
      $x^3 + 2x - 5 = 0$
      Use trial and improvement to find these to 2 d.p.
   (c) Use your graph to find an approximate value for the root of
      $x^3 - 8x + 5 = 0$ in the range −2 to 2.
      Use trial and improvement to find this root to 1 d.p.

(Answers on page 68.)

PROGRESS TEST

 PROGRESS TEST

The following content appears upside down on the page (Progress Test answers):

to 2 of about 0.6. Trial and improvement gives this to 2 d.p. as 0.66

intersection of $y = x^3 + 2x - 5$ and $y = 10x - 10$, giving a very approximate root in the range -2

which is $x^3 + 2x - 5 = 10x - 10$, the roots are found by the x-values of which are the points of

$x^3 - 8x + 5 = 0$ can be re-written as $x^3 + 2x - 5 - (10x - 10) = 0$

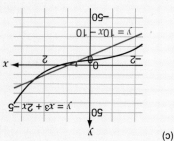

(c)

root to 2 d.p. as 1.33

(b) There is only one root, with very approximate value of 1.3. Trial and improvement gives this

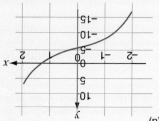

3 (a)

2 (a) -1.8 (b) 3.1 (use the power key on your calculator)

it is what value of x substituted into $x^3 + 5x$ gives a value of 0)

(b) 3.23 ($x^3 + 5x = 50$ can be looked at as $x^3 + 5x - 50 = 0$ or if the former,

1 (a) 1.52

2.8 Inequalities

LEARNING SUMMARY

After studying this section, you will be able to:
- *solve inequalities with one variable*
- *solve inequalities with two variables*

Inequalities in one variable

KEY POINT

≤ means 'less than or equal to'

≥ means 'greater than or equal to'

Inequalities can be shown on number lines:

Whole numbers that fit this inequality are: $-1, 0, 1$

3 is the only whole number that fits

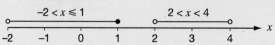

The rules for manipulating inequalities are like those for equations, *except* that **multiplying or dividing each side by a negative number changes the direction of the inequality sign**.

68

Examples

(a) Solve the inequality

$$3(x - 2) > x + 10$$

$$3x - 6 > x + 10$$ ◄── Expand the bracket

$$3x > x + 16$$ ◄── Add 6 to each side

$$2x > 16$$ ◄── Subtract x from each side

$$x > 8$$ ◄── Divide each side by 2

(b) Solve

$$\tfrac{1}{2}(2 - 5a) \leq 11$$

$$2 - 5a \leq 22$$ ◄── Multiply both sides by 2

$$-5a \leq 20$$ ◄── Subtract 2 from each side

$$-a \leq 4$$ ◄── Divide each side by 5

$$a \geq -4$$ ◄── Multiply each side by −1

> Changes the direction of the inequality ($\leq$ to $\geq$).

> When working with a double inequalities start by solving the two inequalities. An integer is any whole number, positive negative or zero.

(c) Find all the possible integer values of n that satisfy $3n + 1 \leq 27 < 5n - 6$.

Splitting up into two inequalities:

$$3n + 1 \leq 27$$
$$3n \leq 26, \ n \leq 8.6666, \text{ so } n = 8, 7, 6, 5, \ldots$$

$$27 < 5n - 6$$
$$33 < 5n, \text{ giving } 5n > 33, \ n > 6.6, \text{ so } n = 7, 8, 9, \ldots$$

Integers that satisfy both are 7 and 8.

When solving inequalities do not forget that multiplying or dividing by a negative number reverses the inequality sign: $-x > 3$, becomes $x < -3$ (multiplying by −1).

Inequalities in two variables

> **KEY POINT**   This is an inequality in two variables: $2x - y > 1$.

> Remember an equation in the form $y = 2x - 1$ has gradient 2 and y-intercept (where it crosses the y-axis) of −1, in other words y = (gradient) x + (y-intercept)

The graph of $2x - y = 1$ (which can be rearranged to $y = 2x - 1$) is a straight line.

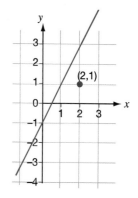

Picking a point under the line, say (2, 1), the value of $2x - y$ is $2(2) - (1) = 3$.
Picking a point above the line, say (1, 2), the value of $2x - y$ is $2(1) - (2) = 0$.
But, by definition, any point on the line, say (1, 1), the value of $2x - y$
is $2(1) - (1) = 1$.

Picking more points supports these findings,
so that summing up:

- On the line: $2x - y = 1$
- Above the line: $2x - y < 1$
- Below the line: $2x - y > 1$

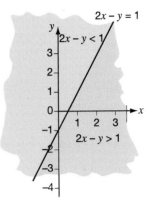

Several inequalities can be shown simultaneously.
The region satisfying $y \leq x$, $x + y \leq 5$ and
$y \geq 1$ is the shaded region here.

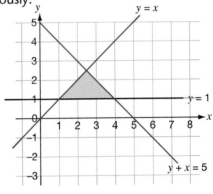

1 Solve these inequalities:
   (a) $x + 7 \leq 3x + 2$ (b) $10 > 2 - 4x$ (c) $2(3x - 7) \geq 5x - 3$
2 List all the integers that satisfy the following inequalities:
   (a) $7 < x \leq 13$ (b) $-6 \leq 3x \leq 6$ (c) $x^2 \leq 25$
3 Write down the four inequalities that define this shaded region.

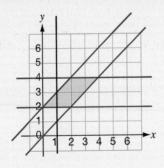

1 (a) $x \geq 2.5$ (if $5 \leq 2x$ then $2x \geq 5$)   (b) $x > -2$ (c) $x \geq 11$
2 (a) 13, 12, 11, 10, 9, 8 (two conditions are $x > 7$ (8, 9, 10 , 11, 12, 13 ,
   and $x \leq 13$ (13, 12, 11, 10, ···)
   (···)
   (b) $-2, -1, 0, 1, 2$
   (c) 5, 4, 3, 2, 1, 0, -1, -2, -3, -4, -5 ($\sqrt{25} = \pm 5$ so $x \leq 5$ and $x \geq -5$)
3 $y \geq 2, y \geq x, y \leq 4, y \leq x + 2$

2.9 Graphs of functions

After studying this section, you will be able to:
- recognise and use the form of the equation for a straight line (linear function)
- draw and recall the graphs of simple non-linear functions

Equations of straight lines

The gradient (slope) of a straight line = $\dfrac{\text{increase in } y\text{-value (vertical change)}}{\text{increase in } x\text{-value (horizontal change)}}$

For an equation of the form $y = mx + c$,

the value of m gives the gradient and c is the value of y where the line cuts the y-axis.

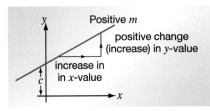

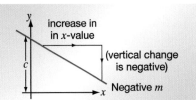

Therefore a straight line of gradient −2 which cuts the y-axis at the point (0, 3) has the equation

$y = -2x + 3$

Examples

(a) Write down the equation of the straight line through the point (0, 3) that is parallel to the line $2y + 3x = 12$.

Re-writing $2y + 3x = 12$ in the form $y = mx + c$ gives $y = -\dfrac{3}{2}x + 6$.

So the line has gradient (value of m) of $-\dfrac{3}{2}$. It has intercept value 6.

Therefore the equation of the line is: $y = -\dfrac{3}{2}x + 6$ or $y = -1.5x + 6$.

(b)

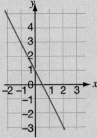

Write down the equation of this straight line.

It has gradient −2, and intercept 1, so has equation $y = -2x + 1$.

KEY POINT
- A straight line with equation $y = mx + c$, has gradient m and intercept c.
- Lines with the same gradient (m) are parallel.
- Multiplying the gradients of two perpendicular lines gives −1.

Graphs of non-linear functions

Sometimes you will be able to see a pattern in the results table. If it breaks down for a point, you may have made a mistake.

First, fill a table of values by substituting into the equation – you may well be asked to fill in a partially complete one.

In the exam you will probably be given a 2 mm grid, possibly with the axes labelled. Always check the values of the units on the x and y-axis scales.

Then draw as smooth a curve as possible through the points – if the smoothness is interrupted by a point, you have probably made a mistake – check your substitution. All the graphs you will be asked to draw will be smooth curves – don't join the points by a series of short straight lines; you will not get any marks.

You need to be able to recognise and draw graphs like these:

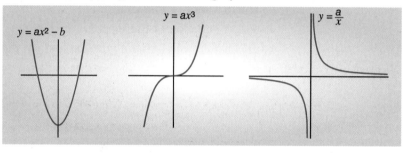

where a and b are positive numbers.

Examples

(a) Draw the graph of $y = x^3 - x + 2$, from $x = -2$ to $x = 2$.

x	−2	−1.5	−1	−0.5	0	0.5	1	1.5	2
y	−4	0.125	2	2.375	2	1.625	2	3.875	8

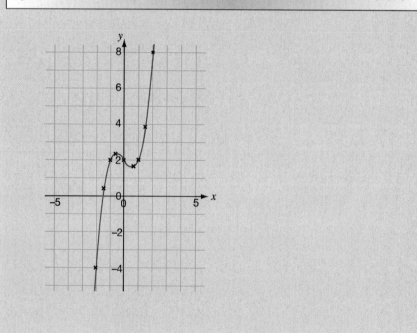

(b) Match the following equations to the graphs:

$y = 2x^2 + 3$

$y = 2x^2 - 3$

$y = -x^2 + 3$

$y = \dfrac{2}{x}$

$y = x^3 - 2x^2$

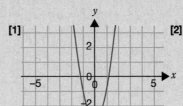
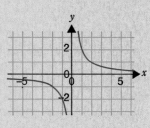

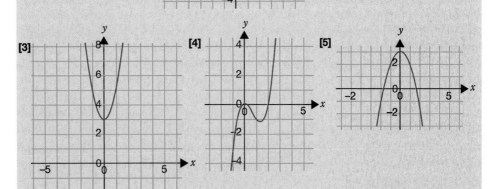

If you are given a numbered grid substitute some 'easy' points such as $x = 0$, this helps identify [1], [3] and [5].

The fact [4] goes through the point (2, 0) fits $y = x^3 - 2x^2$, together with the fact that it has the 'S-shape' characteristic of curves involving **cubic expressions**.

[1] $y = 2x^2 - 3$ [2] $y = \dfrac{2}{x}$ [3] $y = 2x^2 + 3$ [4] $y = x^3 - 2x^2$ [5] $y = -x^2 + 3$

PROGRESS TEST

1 Three lines pass through the point (9, –2). Line [1] is parallel to the x-axis, line [2] is parallel to the y-axis and line [3] is parallel to the line whose equation is $x + y = 99$.

What are the equations of the three lines?

2 Find the equation of the line that passes through the point (0, 4) and is at right-angles to the line $y + 2x = 10$.

3 Draw the graphs of these functions:

(a) $y = 2^x$ from $x = -5$ to $x = 5$.

(b) $y = x^3 + 2x - 1$ from $x = -2$ to $x = 2$.

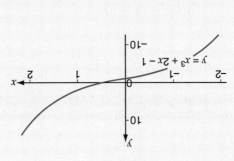

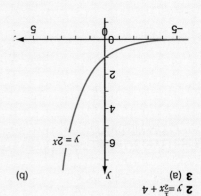

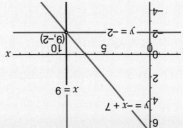

1 [1] $y = -2$, [2] $x = 9$, [3] $y = -x + 7$

2 $y = \frac{1}{2}x + 4$

3 (a)

(b)

2 Algebra

2.10 Sequences

LEARNING SUMMARY

After studying this section, you will be able to:

- generate sequences given a rule and understand some of the vocabulary used in algebra
- construct an algebraic expression for the nth term of a sequence

Generating terms of a sequence

u_1 is shorthand for the first term; t_1 is also used.

Examples

(a) Write down the first three terms of the sequence whose nth term is $n^2 + 1$.

$$u_1 = 1 + 1 = 0 \quad (n = 1)$$
$$u_2 = 4 + 1 = 5 \quad (n = 2)$$
$$u_3 = 9 + 1 = 10 \quad (n = 3)$$

Sequence rules stated in this way are sometimes called 'term to term' rules – one term is expressed in terms of another.

(b) Generate the first three terms of the sequence defined by

$$u_{n+1} = u_n + 3 \text{ with } u_1 = 5.$$
$$u_2 = 5 + 3 = 8$$
$$u_3 = 8 + 3 = 11$$

(c) The nth term of a sequence is $\dfrac{8n}{n+2}$.

(i) Write down the first two terms of the sequence.

(ii) Which term of the sequence has value 6.4?

(i) 1st term $(u_1) = \dfrac{8}{3}$

2nd term $(u_2) = \dfrac{16}{4} = 4$

(ii) $\dfrac{8n}{n+2} = 6.4$

so $8n = 6.4(n + 2)$

$1.6n = 12.8$

giving $n = 8$

Finding the *n*th term

When looking for patterns in sequences look for factors, square and cube numbers and terms going up or down by the same amount.

Look at the sequence: 6, 10, 14, 18, 22, ...

One way to describe it is, 'it starts with 6 and goes up in 4s'.

But this is not very useful if we need to find the 250th term!

A better way is to give the formula for the nth term, $u_n = 4n + 2$.

If the differences between successive terms in a sequence are constant, this difference gives the **coefficient** in the expression for the nth term.

For example the sequence 6, 10, 14, 18, 22 , ... has a constant difference of 4, so the expression for the nth term must involve $4n$. Using just $4n$ generates the sequence 4, 8, 12, 16, 20 , ... which is 2 fewer than the given sequence. So the correct sequence is $4n + 2$.

If there is not a constant difference between terms the expression for the nth term will almost certainly involve an n^2 term.

> In the expression $3x^2 - 2x + 4$ the coefficient of x^2 is 3 and the coefficient of x is -2.

> Differences between terms are 3, 5, 7 and 9, so the nth term probably involves a quadratic.

> Always check that your expression for the nth term agrees with the terms you are given.

> When asked questions like this it is helpful to look at how the pattern is built up.

Examples

(a) Find the nth term of this sequence:

   2, 5, 10, 17, 26, ...

   Squares of whole numbers ($t_n = n^2$) generates
   1, 4, 9, 16, 25, ...

   The corresponding terms are one fewer than the given sequence.
   So the nth term of the given sequence is $u_n = n^2 + 1$.

(b) Here is a sequence of spot patterns.

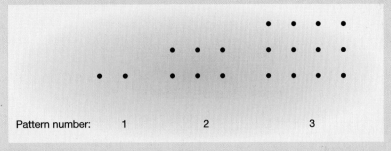

Pattern number: 1   2   3

Find the formula linking the pattern number with the number of spots in each pattern.

   1×2 2×3 3×4

So the nth pattern will have $n(n + 1)$ spots.

PROGRESS TEST

1 Generate these sequences:
   (a) Start with -4 and add 2 each time.
   (b) Start with 10 and subtract 5 each time.
   (c) Write down an expression for the nth term of each of the above sequences.

2 Write down the first four terms of sequences with these nth terms:
   (a) $3n + 2$ (b) $2n(n + 1)$ (c) 3^n

3 Find the nth terms of these sequences:
   (a) 10, 9, 8, 7, ...   (b) 4, 7, 10, ...   (c) 3, 8, 15, 24, ...
   (d) 8, 10, 12, ...   (e) 4, 8, 16, 32, ...

1 (a) $-4, -2, 0, 2, \ldots$ (b) 10, 5, 0, -5, -10, $\ldots$
(c) (i) $u_n = 2n - 6$ (not $2n - 4$ – always check your formula works)
(ii) $u_n = 15 - 5n$ (check your formula works against the given sequence)
2 (a) 5, 8, 11, $\ldots$ (b) 4, 12, 24, $\ldots$ (c) 3, 9, 27, $\ldots$
3 (a) $u_n = 11 - n$ (b) $u_n = 3n + 1$ (c) $u_n = n^2 + 2n$ (d) $u_n = 2n + 6$
(e) $u_n = 2^{n+1}$

2.11 Direct and inverse proportion

After studying this section, you will be able to:
- *set up and solve equations and problems involving proportion*
- *represent and recognise proportional relationship displayed on a graph*

Solving problems involving proportion

KEY POINT

The symbol $\propto$ represents 'proportional to'.

So if y is directly proportional to x, we write $y \propto x$. ('Directly proportional' is used rather than just 'proportional' because there are several different types of proportionality as you will see below.) We can write $y \propto x$ as an equation $y = kx$, where k is a constant (does not depend on x or y).

There are several other types of proportionality:

y is inversely proportional to $x \rightarrow y \propto \dfrac{1}{x}$ or $y = \dfrac{k}{x}$

You need to be able to work with all these.

y is inversely proportional to $x^2 \rightarrow y \propto \dfrac{1}{x^2}$ or $y = \dfrac{k}{x^2}$

y is proportional to $x^2 \rightarrow y \propto x^2$ or $y = kx^2$

y is proportional to $\sqrt{x} \rightarrow y \propto \sqrt{x}$ or $y = k\sqrt{x}$

The value of k depends on the actual problem.
It is usually found by solving the equation formed by substituting for given values of x and y.

Change the proportional statement into an equation involving k. Use the given data to find k, giving an equation linking the two quantities. Solve this equation to find the information asked for.

Examples

(a) The extension of an elastic string (e) is directly proportional to the mass m suspended from it. A mass of 200 g produces an extension of 5 cm. What mass will produce an extension of 3 cm?

$e \propto m$ so $e = km$ where k is a constant.

When $m = 200$, $e = 5$ so $5 = k \times 200$

giving $k = \dfrac{5}{200} = \dfrac{1}{40}$ or $e = \dfrac{1}{40}m$

When $e = 3$ cm, $m = 40 \times 3 = 120$ g

(b) The variable s is inversely proportional to the square of d.
When $s = 8$, the value of d is 5. What is the value of s when $d = 50$?

$s \propto \dfrac{1}{d^2}$ giving $s = \dfrac{k}{d^2}$ ← **Change into an equation**

When $s = 8$, $d = 5$, so $8 = \dfrac{k}{5^2} = \dfrac{k}{25}$, giving $k = 25 \times 8 = 200$ ←── | **Use given values to find k** |

$$s = \frac{200}{d^2}$$

When $d = 50$

$$s = \frac{200}{50^2} = \frac{200}{2500} = 0.08$$ ←── | **Use the equation to find the asked-for quantity** |

Graphs of proportional relationships

> **These are sketch graphs – no scales needed – only the general shape.**

All the proportional relationships mentioned on the previous page can be plotted on a graph, for example if $y \propto x^2$, then $y = kx^2$. The exact form of the graph obviously depends on k, but the general shape is the same whatever the value of k. You need to know the general form of all these graphs of proportional relationships.

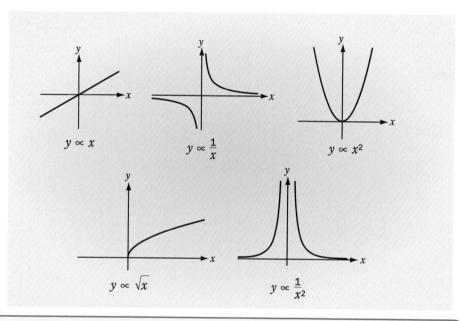

PROGRESS TEST

1 Find the coordinates of A and B. (Hint: what variation does the curve show?)

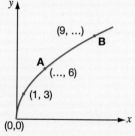

2 (a) y is inversely proportional to x and $y = 10$ when $x = 5$.
Find y when $x = 50$.
(b) y is proportional to the square of x and $y = 40$ when $x = 2$.
Find y when $x = 10$.
(c) y is inversely proportional to x, and $y = 5$ when $x = 2$.
Find x when $y = 1$.

1 A(4, 6), B(9, 9)
2 (a) $y = 1$ (b) $y = 1000$ (c) $x = 10$

2.12 Interpreting graphical information

After studying this section, you will be able to:

● *use and interpret graphs involving practical situations*

Graphs of real-life situations

A very common error made is to assume that a distance–time graph shows a 'picture' of the hill because of its slope. Nothing could be further from the truth – the steeper the hill, all things being equal, the less steep the line on the distance–time graph (slower speed)!

Examples

A cyclist cycles up a hill, has a rest at the top and then freewheels down the other side. This **distance–time graph** 'tells the story' of the journey.

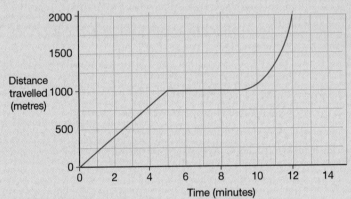

She cycled up the hill at a constant rate (the graph is a straight line), taking 5 minutes to travel the 1000 metres to the top of the hill. She had a rest for 4 minutes (zero distance travelled). When freewheeling down the hill she went faster and faster (the gradient or steepness of the curve increases).

Her speed up this hill was:

$$\frac{\text{distance (metres)}}{\text{time (minutes)}} = \frac{1000}{5} = 200 \, \text{m/min}$$

She took 3 minutes to freewheel down the hill.

Each of these four bowls is filled up with water flowing at a constant rate from a tap.
Which graph goes with which bowl?

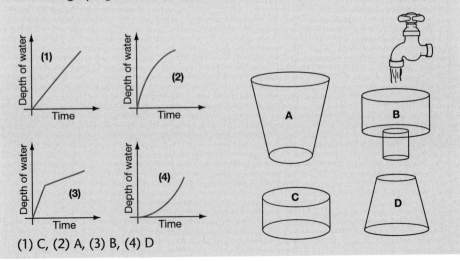

(1) C, (2) A, (3) B, (4) D

> KEY POINT

- The gradient of a **distance–time** graph gives velocity (speed).

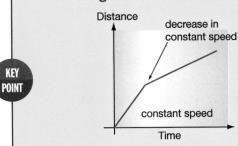

- The gradient of a **velocity** (speed)**–time** graph gives the acceleration.

> PROGRESS TEST

1 Describe each of these journeys in full.

(a)
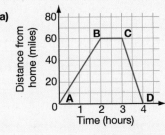

(b)
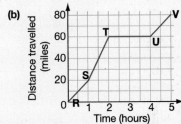

(c) Speed

2 Sketch graphs to show each of these situations:
  (a) The speed of a car stopping at two sets of traffic lights.
  (b) The temperature of a saucepan of water that is heated on a stove and left to boil for a short time before being allowed to cool down.

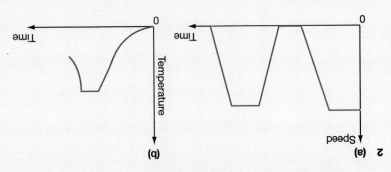

2 (a) Speed / Time
(b) Temperature / Time

1 (a) A → B constant speed of 30 mph, B → C stopped for 1 hour, C → D returned home at 60 mph (note the y-axis is distance *from home, not distance travelled*).
(b) R → S constant speed 20 mph, S → T increase in constant speed to 40 mph, T → U 2 hours stationary, U → V moving with constant speed 20 mph.
(c) E → F steady speed, F → G accelerated steadily, G → H decelerate steadily, H → I constant speed, I → J large deceleration.

2.13 Circles and equations

After studying this section, you will be able to:

● **use and find the equations of circles**

Equation of the circle

KEY POINT

The equation of a circle, centre at the origin and radius *r*, is $x^2 + y^2 = r^2$.

Example

By definition all the points, such as P, on a circle, centre O and radius *r* are distance *r* from the centre.

For a circle radius 6 units, centre (0, 0):
In the shaded triangle, by Pythagoras $6^2 = x^2 + y^2$ but the radius *r* is constant, so $x^2 + y^2 = 6^2$ is an equation that describes all points on the circle with centre (0, 0) and radius 6 units. (As the expression $x^2 + y^2$ involves two squares $x^2 + y^2$ will always be positive.)

(a) Write down the equation of a circle, centre (0, 0) and radius 3 units.

The general equation for a circle, centre (0, 0), and radius *r* is given by:

$$x^2 + y^2 = r^2$$

So the circle with radius 3 units, centre (0, 0) has equation $x^2 + y^2 = 3^2 = 9$.

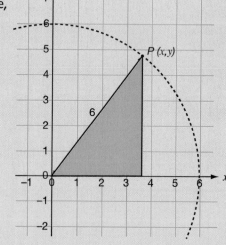

(b) Find the equation of the *x*-coordinates of the points of intersection of the straight line $y = x$ and a circle of radius *r*, centre (0, 0).

The solutions to $x^2 + y^2 = r^2$ and $y = x$ are the values of *x* and *y* where the line $y = x$ cuts the circle $x^2 + y^2 = r^2$.

As $y = x$ then $y^2 = x^2$.

Substituting this into $x^2 + y^2 = r^2$ gives $x^2 + x^2 = r^2$ or $2x^2 = r^2$

so $x = \pm \dfrac{r}{\sqrt{2}}$ and the two points of intersection are

$$\left(\frac{r}{\sqrt{2}}, \frac{r}{\sqrt{2}}\right) \text{ and} \left(\frac{-r}{\sqrt{2}}, \frac{-r}{\sqrt{2}}\right)$$

PROGRESS TEST

1 What is the radius of the circle with equation $x^2 + y^2 = \frac{1}{4}$?

2 Sketch the circle $x^2 + y^2 = 25$, and the straight line $y = x - 1$.
Use algebra to find the points of intersection of the circle and the line.

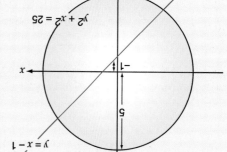

1 $\frac{1}{2}$

2 The circle has radius 5 units. Points of intersection are (4, 3) and (−3, −4).
Need to solve $x^2 + y^2 = 25$ and $y = x - 1$.
$x = 4$ and $y = 4 - 1 = 3$
$x = -3$ and $y = -3 - 1 = -4$
(4, 3) and (−3, −4)

2.14 Transformation of functions and their graphs

LEARNING SUMMARY

After studying this section, you will be able to:
- **understand function notation**
- **apply transformations to the graphs of various functions**

Functions

KEY POINT

$y = f(x)$ is shorthand for 'y is a function of x', for example when

$$y = x^2 + x - 1$$

$$f(x) = x^2 + x - 1$$

and

$$f(x) + 2 = (x^2 + x - 1) + 2$$

$$f(x + 2) = (x + 2)^2 + (x + 2) - 1$$

$$2f(x) = 2(x^2 + x - 1)$$

$$f(2x) = (2x)^2 + (2x) - 1$$

Transforming graphs of functions

What is the connection between the graphs of $y = f(x)$ and $y = f(x) + k$?

For example:

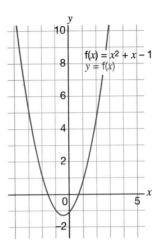 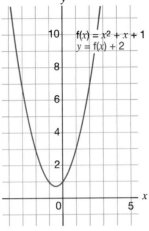

For every value of x, $y = f(x) + k$ will be k more than $y = f(x)$, that is the graph $y = f(x) + k$ is $y = f(x)$ translated by k in the y-direction, i.e. $\begin{pmatrix} 0 \\ k \end{pmatrix}$.

What is the connection between the graphs of $y = f(x)$ and $y = f(x + k)$?

For example:

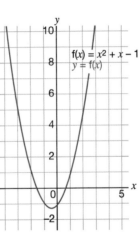 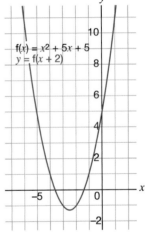

For every value of x, y will take the value it would have had at $x + k$, that is the graph of $y = f(x) + k$ is $y = f(x)$ translated by $-k$ in the x-direction, i.e. $\begin{pmatrix} -k \\ 0 \end{pmatrix}$.

What is the connection between the graphs of $y = f(x)$ and $y = kf(x)$?

For example:

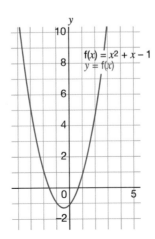 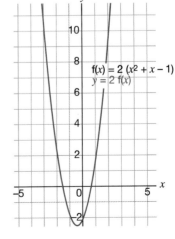

For every value of x, y will be k times bigger, that is the graph of $y = kf(x)$ is $y = f(x)$ stretched by a factor of k in the y-direction.

What is the connection between the graphs of $y = f(x)$ and $y = f(kx)$?
For example:

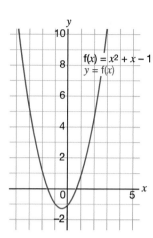

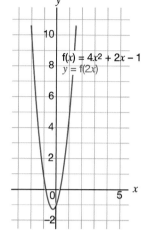

For every value of x, y takes the value it would have had at kx, that is the graph of $y = f(kx)$ is $y = f(x)$ stretched by a factor $\frac{1}{k}$ in the x-direction.

PROGRESS TEST

1 In each case describe the transformation to the first graph to reach the second graph. Sketch the two functions.

(a) $y = x - 2 \rightarrow y = x + 1$ (b) $y = x^2 \rightarrow y = (x - 4)^2$

(c) $y = \sin x° \rightarrow y = \sin (x + 90)°$ (d) $y = x^3 \rightarrow y = (3x)^3$

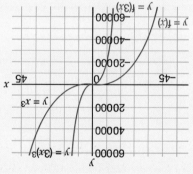

Stretch in the x-direction by a factor $\frac{1}{3}$.

(d)

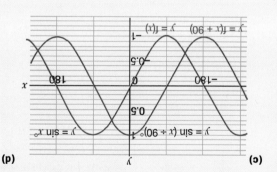

Translation 90° units in the negative x-direction

(c)

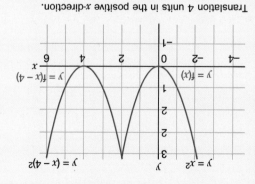

Translation 4 units in the positive x-direction.

(b)

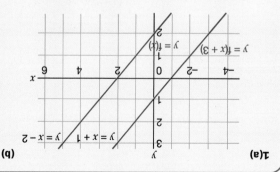

Translation of 3 units in the positive y-direction

1(a)

 **Algebra**

2.15 *Algebraic proof*

After studying this section, you will be able to:
- *understand simple algebraic proof*
- *use algebra to prove simple statements*

Using algebra in proof

Given any precise logical statement, a proof of that statement is a sequence of logically correct steps which shows that the statement is true.

 KEY POINT

In algebraic proof we show that a result is true for x, and providing no arithmetic rules have been broken, it is true for any number subject to the original boundaries set on x – e.g. it must be a positive whole number.

Examples

(a) The nth term in the sequence of triangle numbers 1, 3, 6, 10, 15, ... is given by: $\dfrac{n}{2}(n + 1)$. Prove that eight times any triangle number is one less than a square number.

If T is a triangle number then we need to prove that $8T + 1$ is a square number.

$8T + 1$ is given by $4n(n + 1) + 1$ which simplifies to $4n^2 + 4n + 1$.
But $4n^2 + 4n + 1 = (2n + 1)^2$ which is a square number – we have proved the result.

(b) The nth term of the so-called rectangle numbers is $n(n + 1)$.
Prove that rectangle numbers are always even.

We need to prove that, for positive integers n, $n(n + 1)$ is always even.
If n is even then $(n + 1)$ is odd, but (even) × (odd) is always even.
If n is odd then $(n + 1)$ is even, but (odd) × (even) is always even.
So rectangle numbers are always even.

(c) Prove that if you add two consecutive rectangle numbers ($u_n = n(n + 1)$) and half the answer, the result is always a square number.

$$u_n = n(n + 1) = n^2 + n$$

and the next term is

$$u_{n+1} = (n + 1)(n + 2) = n^2 + 3n + 2$$

so

$$u_n + u_{n+1} = 2n^2 + 4n + 2$$
$$= 2(n^2 + 2n + 1)$$

Half of this is $= n^2 + 2n + 1$.

But this can be written as $(n + 1)^2$, which is a square number – the result we wanted.

PROGRESS TEST

1 Prove that the sum of any three consecutive numbers is divisible by 3.

2 Prove that the sum of four consecutive numbers is not divisible by 4.

3 Prove that for any three consecutive numbers, the difference between the squares of the first and last numbers is 4 times the middle number.

1 $n + (n + 1) + (n + 2) = 3n + 3 = 3(n + 1)$

2 $n + (n + 1) + (n + 2) + (n + 3) = 4n + 6 = 4(n + 1) + 2$

3 If numbers are (n), $(n + 1)$, $(n + 2)$ then $(n + 2)^2 - (n)^2 = n^2 + 4n + 4 - n^2 = 4n + 4 = 4(n + 1)$, which is 4 times the middle number $(n + 1)$.

Sample GCSE questions

1 (a) Expand and simplify the following expressions:
   (i) $3(c + 3) + 5(2x - 1)$
   (ii) $x(x^2 - 5)$ **[4]**
   (b) Factorise these expressions:
   (i) $4x^2 + 8xy$
   (ii) $x^2 - 16$ **[3]**
   (c) (i) Factorise $x^2 - 7x + 6$
   (ii) Hence simplify $\dfrac{x^2 - 7x + 6}{x^2 - 36}$ **[5]**

(a) (i) $3(c + 3) + 5(2x - 1)$ ← Expand brackets ✔
        $= 3c + 9 + 10x - 5$
        $= 3c + 10x + 4$ ← Collect like terms ✔

   (ii) $x(x^2 - 5)$ ← Expand the bracket ✔✔
        $= x^3 - 5x$

        Don't forget to factorise fully.

(b) (i) $4x^2 + 8xy = 4x(x + 2y)$ ✔✔

   (ii) $x^2 - 16 = (x - 4)(x + 4)$ ✔

        You need to be able to recognise the difference of two squares.

(c) (i) $x^2 - 7x + 6 = (x - 6)(x - 1)$ ← Need two numbers whose product is 6 and sum −7 ✔✔

   (ii) $\dfrac{x^2 - 7x + 6}{x^2 - 36} = \dfrac{(x - 6)\,(x - 1)}{x^2 - 36}$ ← Use part (i) ✔

        $= \dfrac{(x - 6)\,(x - 1)}{(x - 6)\,(x + 6)}$ ← Factorise the denominator ✔

        $= \dfrac{(x - 1)}{(x + 6)}$ ← Divide/cancel top and bottom by $(x - 6)$ ✔

2 The sum of a number and its reciprocal is 2.05.
Set up and solve an equation to find the two numbers. **[6]**

Let x be the number.

So $x + \dfrac{1}{x} = 2.05$ ✔

   Always say what your unknown represents when setting up an equation.

Giving $x^2 + 1 = 2.05x$ ← Multiply through by x ✔

or $x^2 - 2.05x + 1 = 0$ ← Put into the usual quadratic form

   Unless you can spot the factors, use the formula – but put what values you are using for a, b and c – there may be method marks.

Using the formula $x = \dfrac{-b \pm \sqrt{b^2 - 4ac}}{2a}$ *with* $a = 1, b = -2.05, c = 1$ ✔

   $= \dfrac{2.05 \pm \sqrt{2.05^2 - 4}}{2}$ ✔

giving $x = 1.25$ *or* 0.8, *so the number is 1.25 or 0.8* ✔✔

   Check to make sure that the numbers fit the condition given in the question.

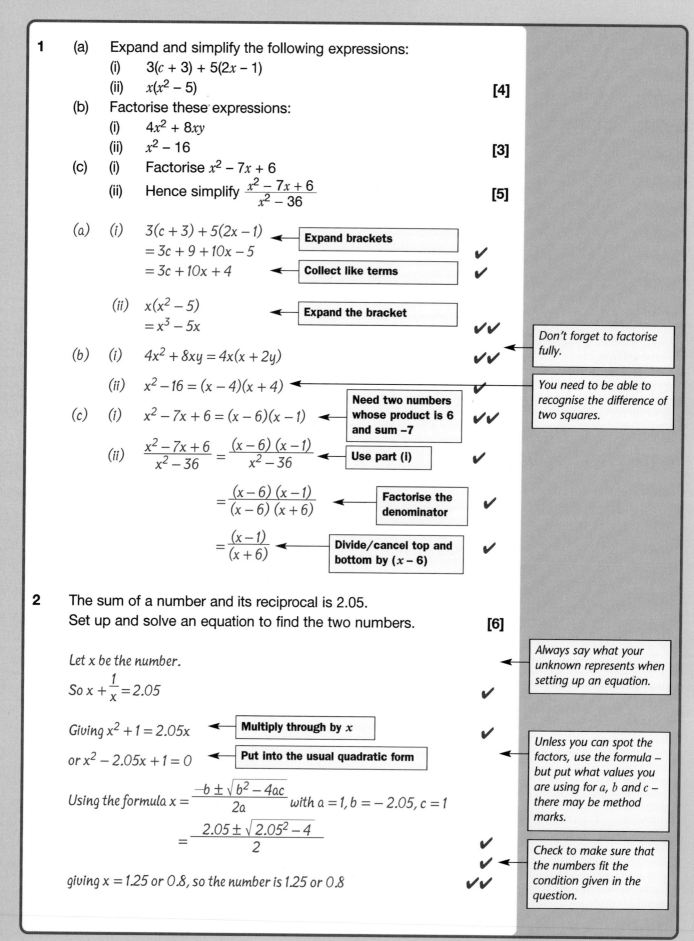

Sample GCSE questions

3 Solve the equations
   (a) $5x + 17 = 3(x + 6)$ **[3]**

   (b) $\dfrac{2x - 3}{6} + \dfrac{x + 2}{3} = \dfrac{5}{2}$ **[4]**

   (a) $5x + 17 = 3(x + 6)$ ◄── Expand the bracket
        $5x + 17 = 3x + 18$ ✔

        $2x = 1$ ◄── Collect terms   ✔

        $x = 0.5$ ✔

   (b) $\dfrac{2x - 3}{6} + \dfrac{x + 2}{3} = \dfrac{5}{2}$ ✔

        $\dfrac{6(2x - 3)}{6} + \dfrac{6(x + 2)}{3} = \dfrac{6(5)}{2}$ ◄── Multiply through by 6 to remove the fractions

        $(2x - 3) + 2(x + 2) = 3 \times 5$ ✔
        $2x - 3 + 2x + 4 = 15$ ◄── Expand the brackets   ✔

        $4x = 14$
        $x = 3.5$ ◄── Check by substituting back   ✔

4 Here is the graph of the function

   $$y = x^3 + x - 3$$

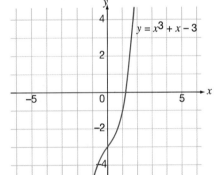

   Use trial and improvement to find one root of the equation:

   $$x^3 + x - 3 = 0$$

   Give your answer to 2 d.p.
   Show all your trials.   **[4]**

   From the graph the solution must be round about 1.2   ✔

x	$x^3 + x - 3$ (target = 0)
1.2	−0.072
1.3 ✔	0.497
1.25	0.203 125
1.23	0.090 867
1.22	0.035 848
1.21	−0.018 439
1.215 ✔	0.008 613 375

   ◄── Solution is between 1.2 and 1.3

   ◄── Solution is between 1.21 and 1.22

   ◄── Solution is between 1.21 and 1.215

   Solution of $x^3 + x - 3 = 0$ is 1.21 to 2 d.p.

Sample GCSE questions

5   Solve these simultaneous equations algebraically

$3x + 2y = 7$

$2x + 5y = 1$ **[4]**

$3x + 2y = 7 \: [1]$

$2x + 5y = 1 \: [2]$

$[1] \times 5 - [2] \times 2$

$\quad 15x + 10y = 35$ ← Solve by elimination   ✔

$\quad -4x - 10y = -2$ ✔

$\qquad\quad 11x = 33$

$\qquad\qquad x = 3$ ✔

Substituting in [1] for x: $9 + 2y = 7$

$y = -1$ ✔

Label the equations and clearly state what you are doing.

6   List all the integers that fit this inequality:

$8 < 3n \leq 21$ **[3]**

$8 < 3n \leq 21$

$8 < 3n$

So $n = 3, 4, 5, \ldots$ ← Find solutions by treating each condition separately   ✔

$3n \leq 21$ so $n = 7, 6, 5, 4, 3, 2, \ldots$ ✔

so $n = 3, 4, 5, 6, 7$ ✔

7   **(a)**   Find the equation of the straight line that passes through the points (1, 1) and (3, 5).   **[3]**

(b)   Find the equation of the straight line that is at right-angles to the line $y = x + 1$ and that passes through the point (4, 0).   **[3]**

(a)

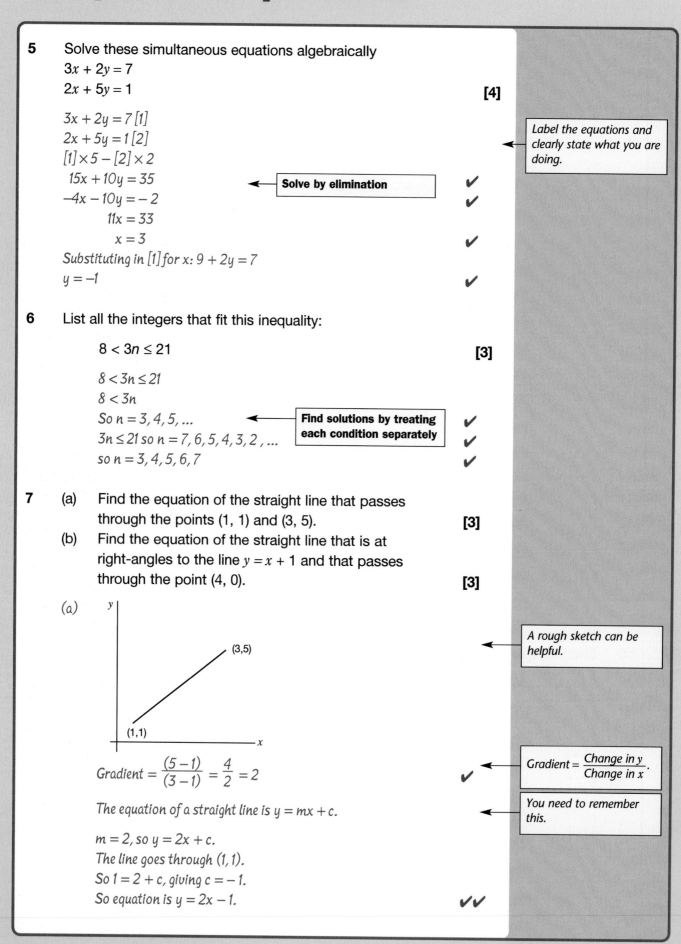

A rough sketch can be helpful.

Gradient $= \dfrac{(5-1)}{(3-1)} = \dfrac{4}{2} = 2$ ✔

Gradient $= \dfrac{\text{Change in } y}{\text{Change in } x}$.

The equation of a straight line is $y = mx + c$.

You need to remember this.

$m = 2$, so $y = 2x + c$.

The line goes through (1, 1).

So $1 = 2 + c$, giving $c = -1$.

So equation is $y = 2x - 1$.   ✔✔

Sample GCSE questions

(b)   When two straight lines cut at right angles the product of their gradients is −1.

> *You need to remember this.*

Gradient of line = −1 (−1 × 1 = −1) ✔
General straight line equation is y = mx + c, with m = −1.
The line goes through (4, 0) so 0 = −4 + c.
So c = 4
So equation is y = −x + 4 ✔✔

8   (a)   Write the values of the following expressions in descending order of size, when $x > 1$. **[2]**

$$x^{-1} \quad x^2 \quad \sqrt{x} \quad x^{-2}$$

(b)   What happens to the expressions when $x = 1$? **[1]**

> *$\frac{1}{x}$ is always less than 1 when $x > 1$ (and positive), a positive number less than one squared is less than the number itself.*

(a) $x^2 \quad \sqrt{x} \quad x^{-1} \quad x^{-2}$ ✔✔

(b)   When x = 1 all are equal to 1. ✔

9   A rectangle is 3 cm longer than it is wide.
The area is 108 cm².
Use algebra to find its length. **[4]**

Let x = the width of the rectangle.

> *Always state what your unknowns represent.*

Area of rectangle = x(x + 3) = x² + 3x ✔
$$x^2 + 3x = 108$$
$$x^2 + 3x - 108 = 0$$ ✔
$$(x + 12)(x - 9) = 0$$ ✔
$$x = 9 - \text{so width is } 9\,cm$$ ✔

> *A rough sketch can be helpful.*

> *The negative solution has no meaning in this situation.*

10   The curve $y = x^2 + 2$ and the straight line $y = x + 2$ intersect at two points.

Find the coordinates of these two points. **[4]**

> *The question is really asking for the solutions to these two simultaneous equations, which is where the curve and straight line intersect.*

Substituting y = x + 2 into y = x² + 2 ✔
gives x + 2 = x² + 2
So x² − x = 0 ✔
x(x − 1) = 0, giving x = 0 or x = 1 ✔
So y = 2 or y = 3
The points of intersection are (0, 2) and (1, 3) ✔

 **Algebra**

Sample GCSE questions

11 Make n the subject of the formula $a = \dfrac{4(n-b)}{n}$. **[4]**

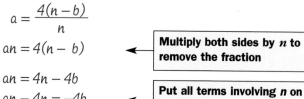

$$a = \frac{4(n-b)}{n}$$

$$an = 4(n-b)$$

Multiply both sides by n to remove the fraction ✔

$$an = 4n - 4b$$

$$an - 4n = -4b$$

Put all terms involving n on the left ✔

$$n(a-4) = -4b$$

Using brackets makes the algebra much easier to see ✔

So $n = \dfrac{-4b}{a-4}$ or $\dfrac{4b}{4-a}$

Dividing both sides by $(a-4)$, either answer is acceptable, multiplying top and bottom by -1 gives a positive numerator. ✔

12 Here is a sketch of the function $y = x^2 + 2$.

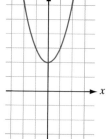

Sketch the functions:
(a) $y = x^2 - 2$
(b) $y = (x+2)^2 + 2$
(c) $y = -(x^2+2)$ **[3]** ✔✔✔

You need to be able to recall these transformations of the graph $y = f(x)$: $y = f(x+a)$, $y = af(x)$, $y = f(a)+a$ and $y = f(ax)$ – some of them are not obvious.

(a)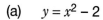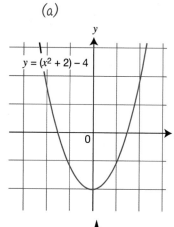
$y = (x^2+2) - 4$

(b)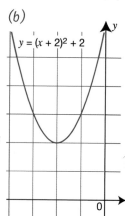
$y = (x+2)^2 + 2$

(c)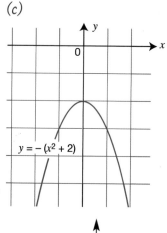
$y = -(x^2+2)$

$y = (x^2+2) - 4$

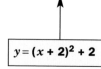
$y = (x+2)^2 + 2$

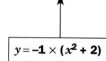
$y = -1 \times (x^2+2)$

Sample GCSE questions

13 (a) Given that $\dfrac{x}{2x-3} = \dfrac{3}{x}$ show that $x^2 - 6x + 9 = 0$ **[3]**

(b) Solve the quadratic equation $x^2 - 6x + 9 = 0$. **[2]**

(a) $\dfrac{x}{(2x-3)} = \dfrac{3}{x}$ ← *Brackets help show the grouping of terms.*

$x^2 = 3(2x - 3)$ ✔

$x^2 = 6x - 9$ ✔

$x^2 - 6x + 9 = 0$ ✔

(b) $x^2 - 6x + 9 = 0$

$(x - 3)(x - 3) = 0$ ✔

so $x = 3$ ✔

> *With 'show that' questions, always show clearly exactly what you have done – otherwise the examiners may think you are trying to fool them.*

> *A useful technique when given two fractions equal to each other is to 'cross-multiply', e.g. if $\frac{A}{B} = \frac{R}{S}$ then $AS = BR$: this works as a result of multiplying both sides by B and then the result by S.*

> *If it can be solved by factorisation we need two numbers whose sum is –6 and product 9 – these are –3 and –3.*

14 A function f(x) is given by f(x) = $x^3 - 2x + 1$. Complete this table.

x	–2	–1	–0.5	0	0.5	1	1.5	2
f(x)	–3			1				5

> *Always check the results in a table – a slip in the table may result in a wrong graph.*

Draw a graph of the function, using a scale of 1 cm per unit.
By drawing a suitable line find the solutions of the equation
$x^3 - 2x + 1 = x + 1$. **[6]**

x	–2	–1	–0.5	0	0.5	1	1.5	2
f(x)	–3	2	1.875	1	0.125	0	1.375	5

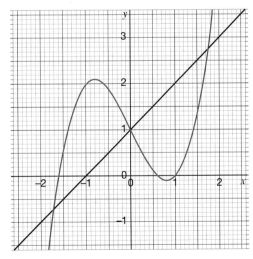

✔✔

> *Make sure your points are joined up to make a smooth curve.*

✔✔

> *Solutions to the equation $x^3 - 2x + 1 = x + 1$ are given by the x-coordinates of the points of intersection of $y = x^3 - 2x + 1$ and $y = x + 1$.*

There are three solutions:

0, –1.7 and 1.7 ✔✔

Sample GCSE questions

15 Here is a number pattern that is described by arrangements of dots. They could be called 'flag numbers', so the 1st flag number is 2, the 2nd flag number is 6, the 3rd 12 and so on.

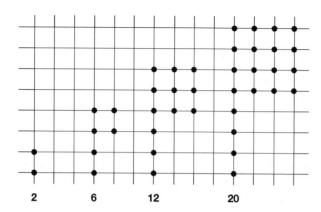

2   6   12   20

(a) (i) Write down the 5th flag number, without drawing a pattern of dots.

  (ii) Describe how you worked out your answer to (i). **[2]**

(b) Write down an expression for the *n*th flag number. **[2]**

(c) Use your answer in (b) to explain why flag numbers are always even. **[2]**

(a) (i) $5^2 + 5 = 30$ ✔✔

  (ii) *Squaring five and adding five to the result* ✔

(b) $n^2 + n$ *or* $n(n + 1)$ ✔✔

(c) $n(n + 1)$ *is always be even because if n is odd, then* $(n + 1)$ *is even and odd × even is always even, and if n is even, then* $(n + 1)$ *is odd and even × odd is even.* ✔✔

> *Get an insight into the sequence by looking at the shapes that built it up. For this one the flag itself is a square n by n spots with a pole n spots high.*

16 Simplify:

(a) $14a^6 \div 2a^2$

(b) $c^2d^{-2} \times c^{-3}d$ **[3]**

(c) $(p^{-2})^{-3}$

(d) $(16m^8)^{\frac{1}{2}}$ **[3]**

(a) $14a^6 \div 2a^2 = 7a^4$ ✔

(b) $c^6d^{-2} \times c^{-3}d = c^{2-3} \times d^{-2-1} = c^{-1}d^{-1}$ ✔✔

(c) $(p^{-2})^{-3} = p^{(-2 \times -3)} = p^6$ ✔

(d) $(16m^8)^{\frac{1}{2}} = \sqrt{16m^8} = \sqrt{16} \times \sqrt{m^8} = 4m^4$ ✔✔

Exam practice questions

1 Solve these equations:

    (a) $4a - 7 = -20$ **[2]**

    (b) $5(2x - 1) + 6x = 7 - 8x$ **[3]**

    (c) $\dfrac{2}{x} = 4$ **[2]**

2 Solve algebraically the simultaneous equations:

    $3x - 2y = 10$

    $5x + 4y = 2$

    Show your working.   **[3]**

3   (a)   The length of a string and the frequency of the note it produces are inversely proportional. A string of length 65 cm is tuned to D, which has a frequency of 147 hertz (Hz).
Calculate the length of string that would produce a note of frequency 110 Hz.   **[2]**

    (b)   For a string of fixed length the frequency of the note produced is directly proportional to the square root of the tension in the string. For a particular tension, the note produced is 196 Hz.
What frequency note is produced when this tension is doubled?   **[2]**

4 This is a drawing of the circle whose equation is $x^2 + y^2 = 25$ and a straight line.

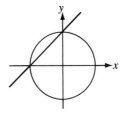

    (a)   Write down the equation of the straight line.   **[2]**

    (b)   Use algebra to find the points of intersection of the circle and the straight line $y = x + 1$.   **[5]**

5 Given that n is an integer, write down the next integer after n.
Show that the difference between two consecutive integers squared is always an odd number, for example $5^2 - 4^2 = 25 - 16 = 9$ which is an odd number.   **[4]**

Exam practice questions

6   This sketch shows the graph of $y = f(x)$.

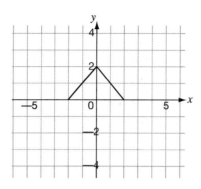

Sketch the graphs of:

(a) $y = f(x) + 1$ [1]

(b) $y = f(x - 4)$ [1]

(c) $y = \frac{1}{2} f(x)$ [1]

7   (a)   Find the values of a and b when
           $x^2 + 6x + 11 \equiv (x + a)^2 + b$ [3]

   (b)   Find the nth term of the following sequences:

           (i)   1, 4, 9, 16, 25, 36, …

           (ii)   0, 3, 8, 15, 24, 35, …

           (iii)   2, 6, 12, 20, 30, 42, …

           (iv)   3, 6, 12, 24, 48, … [5]

8   Solve: $\dfrac{3}{2x - 1} + \dfrac{2}{x + 2} = 1$ [7]

9   The nth hexagon number is given by $n(2n - 1)$.

   What hexagonal number is 45? [4]

3 Shape, space and measures

Overview

Topic	Section	Studied in class	Revised	Practice questions
3.1 Properties of triangles, quadrilaterals and polygons	The basic angle facts			
	Angle facts of triangles			
	Angles in quadrilaterals and polygons			
	Properties of quadrilaterals			
3.2 Pythagoras' theorem	Using Pythagoras' theorem			
	Problems in three dimensions			
3.3 Trigonometry	Trigonometry in right-angled triangles			
	Area of a triangle			
	Sine, cosine and tangent of any angle			
	Sine and cosine rules			
3.4 Congruent triangles	Congruent triangles			
3.5 Circles	Parts of a circle			
	Cyclic quadrilaterals			
	Angle facts involving the circle			
3.6 Proof	Using proof in geometry			
3.7 Transformations	Reflection			
	Translation			
	Rotation			
	Enlargement			
	Combined transformations			
3.8 Coordinates in three dimensions	2D coordinates			
	3D coordinates			
3.9 Vectors	Writing vectors			
	Adding and subtracting vectors			
	Multiplying vectors			
3.10 Constructions and loci	Constructing triangle *ABC* from given information			
	The perpendicular bisector of a line			
	The perpendicular from a point to a line			
	To bisect an angle			
	Loci			
3.11 Dimensions, areas and volumes	Dimensions			
	Areas and volumes of similar figures			
3.12 Volumes of prisms, cones, pyramids and spheres	Prism			
	Cone			
	Pyramid			
	Sphere			
3.13 Circles and spheres	The sphere			
	Arcs, sectors and segments			

3.1 Properties of triangles, quadrilaterals and polygons

> **LEARNING SUMMARY**
>
> *After studying this section, you will be able to:*
> - *use the basic angle facts involving straight lines and parallel lines*
> - *use the angle facts connected with triangles, quadrilaterals and polygons*
> - *identify quadrilaterals by their geometric properties*

The basic angle facts

You must remember the basic angle facts such as angles on a straight line sum to 180°, recognise acute, right, obtuse and reflex angles.

You must also remember the parallel line angle facts.

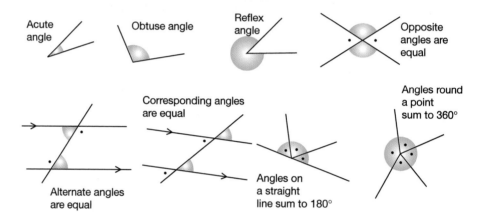

Angle facts of triangles

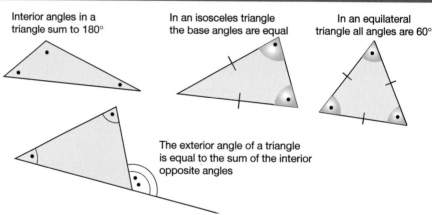

Angles in quadrilaterals and polygons

The sum of the **interior angles** in a quadrilateral is 360°.

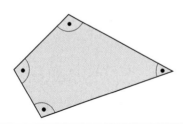

The sum of the **interior angles** of an *n*-sided polygon is $(n - 2) \times 180°$.

The sum of the interior angles of this octagon is:

$$(8 - 2) \times 180° = 1080°$$

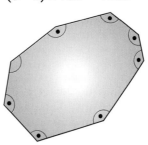

The sum of the **exterior** angles of any polygon is 360°.
So the sum of these marked angles is 360°.

> The relationship between exterior and interior angle is simple: exterior + interior = 180° – they are angles on a straight line.

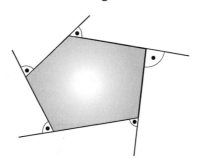

Examples

(a) Find the size of the angles marked with letters.

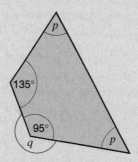

> You must give a reason for each of your steps where possible, and lay your work out logically and clearly.

angle $q = 360° - 95° = 265°$ ◄—— **Angles round a point sum to 360°**

$2p + 135° + 95° = 360°$ ◄——

so $p = (360 - 135 - 95)° \div 2$ | **Internal angles of a quadrilateral sum to 360°**

$\quad = 65°$

(b) Two interior angles of a polygon are 140° and 120°.
Each of the remaining interior angles is 128°.
How many sides has the polygon?

The exterior angles of the first two angles are $180° - 140° = 40°$ and $180° - 120° = 60°$.

The other exterior angles are $180° - 128° = 52°$.

> The fact that the sum of exterior angles of any polygon is 360° can quite often turn a hard problem into a simple one.

The total of the first two exterior angles is $360° - 40° - 60° = 260°$ (sum of exterior angles in any polygon is 360°). Each of the remaining ones is 52° – so there are $260 \div 52 = 5$ of them.

This means that the polygon has $5 + 2 = 7$ sides.

Properties of quadrilaterals

You need to be able to recall these geometric properties of these quadrilaterals.

Square

- all sides equal and opposite sides parallel
- all angles 90°
- four lines of symmetry
- rotation symmetry order 4
- diagonals bisect at right angles

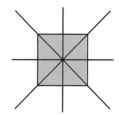

Rectangle

- opposite sides equal and parallel
- all angles 90°
- rotation symmetry order 2
- diagonals bisect at right angles

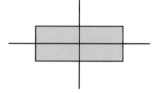

Parallelogram

- opposite sides equal and parallel
- opposite angles equal
- no lines of symmetry
- rotation symmetry order 2

Rhombus

- all sides equal
- opposite sides parallel
- two lines of symmetry
- rotation symmetry order 2
- diagonals bisect at right angles

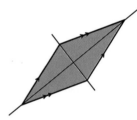

Kite

- one line of symmetry
- diagonals intersect at right angles
- two pairs of adjacent sides equal

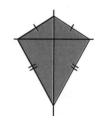

Trapezium

- one pair of sides parallel

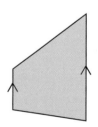

Arrowhead

- one line of symmetry
- two pairs of adjacent sides equal

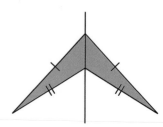

1 The sum of the interior angles of a polygon is 900°.
How many sides does it have?

2 Find the size of the angles marked a, b and c.

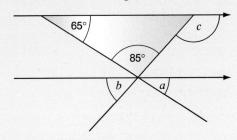

3 *PQRS* is an arrowhead. $PQ = QR$ and $PS = SR$.
Find the size of the angles marked with letters.

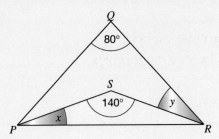

3.2 Pythagoras' theorem

After studying this section, you will be able to:

- **use Pythagoras' theorem to find the third side in a right-angled triangle**
- **find the length of a line joining two points on a graph**
- **use Pythagoras' theorem to solve problems in three dimensions**

Using Pythagoras' theorem

Make sure you remember that the hypotenuse is the side opposite the right angle.

KEY POINT

Pythagoras' theorem is usually stated as:

In a right-angled triangle, the square on the hypotenuse = the sum of the squares on the other two sides

or $a^2 = b^2 + c^2$

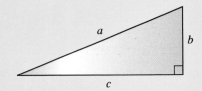

Examples

(a) Two joggers run 8 km north and then 5 km west. What is the shortest distance, to the nearest tenth of a km, they must travel to return to their starting point?

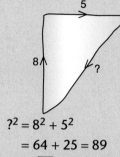

$$?^2 = 8^2 + 5^2$$
$$= 64 + 25 = 89$$
so $? = \sqrt{89} = 9.4$ km to nearest 0.1.

(b) In a TV catalogue, a TV screen is listed as being 40 inches, the diagonal distance across the screen. The screen measures 19.5 inches in height. What is the actual width of the screen to the nearest inch?

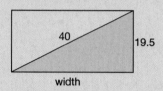

width

$$40^2 = 19.5^2 + \text{width}^2$$
so $\text{width}^2 = 40^2 - 19.5^2 = 1600 - 380.25$
$$= 1219.75$$
giving width $= \sqrt{1219.75} = 35$ inches to the nearest inch.

The length of the line joining two points on a graph

As the standard coordinate grid is literally made up from squares, there are plenty of right-angled triangles formed.

Example

There are two points on a grid, A (5, 5) and B (17, 20).
Calculate: (a) the distance of A from the origin and
(b) the length of the line AB.

(a) $OA^2 = 5^2 + 5^2 = 25 + 25 = 50$
so $OA = \sqrt{50} = 7.07$ to 3 s.f.

(b) $AB^2 = 12^2 + 15^2 = 144 + 225 = 369$
so $AB = \sqrt{369} = 19.2$ to 3 s.f.

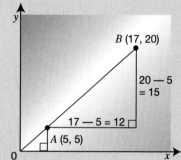

Problems in three dimensions

KEY POINT

Any drawings of 3D situations will be in perspective, so a right-angled triangle may not look like one. It is a very good idea to make a separate 2D sketch of the triangle you are using.

Example

This diagram shows a cuboid 5 cm by 3 cm by 2 cm.

Calculate the lengths DB and BH.

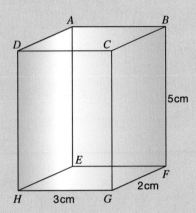

In triangle DCB:

$DB^2 = 3^2 + 2^2$

$= 9 + 4 = 13$

so $DB = 3.61$ cm to 3 s.f.

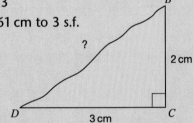

In triangle DBH:

$BH^2 = DH^2 + DB^2$

$= 25 + 13 = 38$

$BH = 6.2$ cm to 2 s.f.

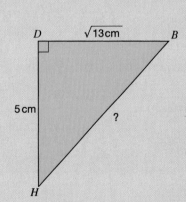

1 A rectangle is 7 cm by 8 cm. How long are its diagonals?

2 The diagonals of a rhombus are 10 cm and 13 cm.
What is the length of the side of the rhombus?

3 The sides of a triangle are 7, 11 and 12 units.
Is it a right-angled triangle?

4 Calculate the distance between these two points:
(a) (1, 4) and (5, 7)   (b) (4, 5) and (−4, −5)

5 Find the lengths of the lettered sides.
Give your answers to 3 s.f.

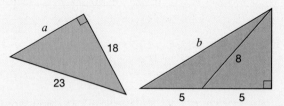
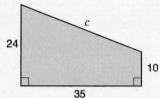

6 *ABCD* is a square-based pyramid, with the apex *E* directly over the corner *A*.
The square base has sides 5 m long and *AE* is 10 m.
Calculate (a) *EB* and (b) *EC*.

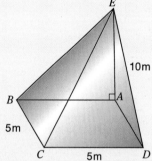

7 *ABFE* is a horizontal plane, *AB*, *EF* and *CD* are horizontal lines of length 50 m,
BF is 12 m and *FD* and *CE* are vertical lines of length 4 m.
Find (a) *BD* and (b) *CB*.

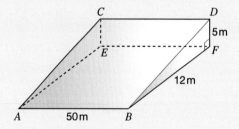

PROGRESS
TEST

1 $\sqrt{113}$ = 10.6 cm (3 s.f.)
2 $\sqrt{5^2 + 6.5^2}$ = 8.20 cm (3 s.f.)
3 No – because 12^2 = 144 is not 49 + 121 (= 170)
4 (a) 5   (b) 12.8 (3 s.f.)
5 (a) 14.3   (b) 11.8   (c) 37.7
6 (a) 11.2 m   (b) 12.2 m
7 *BD* = 13 m, *CB* = 52 m

3.3 Trigonometry

LEARNING SUMMARY

After studying this section, you will be able to:
- *use the sine, cosine and tangent ratios to find angles and sides in right-angled triangles*
- *find the area of a triangle using the formula $A = \frac{1}{2}ab \sin C$*
- *find the sine (sin), cosine (cos) and tangent (tan) of any angle*
- *use the sine and cosine rules*

Trigonometry in right-angled triangles

The equations of trigonometry connect together the sides and angles of a right-angled triangle.

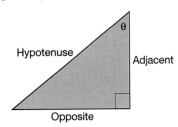

Sides are referenced to an angle. Opposite means opposite the angle and adjacent means next to the angle – but not the hypotenuse.

KEY POINT

You need to learn these three ratios.

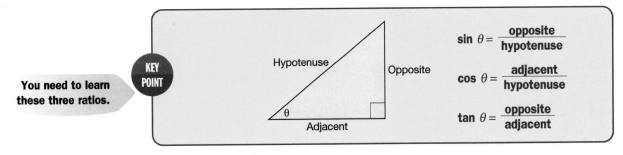

$$\sin \theta = \frac{\text{opposite}}{\text{hypotenuse}}$$

$$\cos \theta = \frac{\text{adjacent}}{\text{hypotenuse}}$$

$$\tan \theta = \frac{\text{opposite}}{\text{adjacent}}$$

Examples

(a) A cliff railway climbs at an angle of 50° to the horizontal. The track is 130 m long. Work out the height of the cliff.

$$\sin 50 = \frac{\text{height}}{130}, \text{ so height} = 130 \times \sin 50$$

Don't forget to make a rough sketch of the situation.

$$= 130 \times 0.7660 \dots$$
$$= 99.6 \text{ m to 3 s.f.}$$

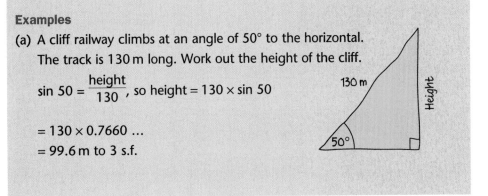

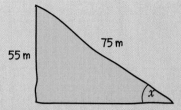

(b) A kite, on a string 75 m long, is 55 m high. What angle does it make to the horizontal?

> You know the hypotenuse and opposite side and need to find the angle, so you need to use sine.

$$\sin x = \frac{55}{77}, \text{ so } x = \sin^{-1}\left(\frac{55}{77}\right) = 47° \text{ (2 s.f.)}$$

This means *the angle whose sine is $\frac{55}{75}$* – you need to use the inverse sine function to find the actual angle, using the inverse key and the sin key or the sin^{-1} key, depending on your calculator.

> Angles of elevation (and depression) are always measured from the horizontal.

(c) The angle of elevation of the top of an aerial from a point 24 m from the base is 55°.
Find the height of the aerial.

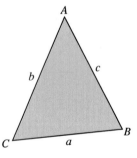

> We know the angle and its adjacent side and want to find the opposite side, so the tan ratio is the one to be used.

$$\tan 55 = \frac{\text{height}}{24}$$

so height $= 24 \times \tan 55$ ← Multiply both sides of the equation by 24

$= 34.3 \text{ m (3 s.f.)}$

Area of a triangle

There is a shorthand method to identify angles and sides of a triangle: angle A is labelled as that angle opposite side a and so on. Angles are shown with capital letters and sides with lower case letters.

> **KEY POINT**
>
> For any triangle, if you know one angle (say A) and its two adjacent sides (b and c):
>
> The area of the triangle is given by $\frac{1}{2}ab \sin C$ (or $\frac{1}{2}ac \sin B$ or $\frac{1}{2}bc \sin A$).

Example
Find the area of this triangle.

Area $= \frac{1}{2}ab \sin C$

$= \frac{1}{2} \times 14 \times 12 \times \sin 65$

$= 76.1 \text{ cm}^2 \text{ (to 3 s.f.)}$

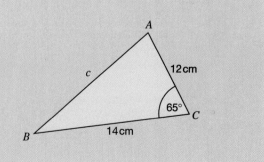

Sine, cosine and tangent of any angle

This diagram shows a circle with radius 1 unit, with centre at the origin. The point P, which has coordinates (x, y), can be anywhere on the circumference.

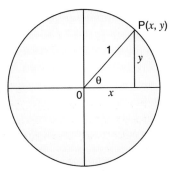

From your work in the section 'Trigonometry in right-angled triangles', check that you agree with these:

$$\sin \theta = \left(\frac{\text{opposite}}{\text{hypotenuse}}\right) = \frac{y}{1}, \quad \cos \theta = \left(\frac{\text{adjacent}}{\text{hypotenuse}}\right) = \frac{x}{1}, \quad \tan \theta = \left(\frac{\text{opposite}}{\text{adjacent}}\right) = \frac{y}{x}.$$

The circle can be divided up into quadrants like this:

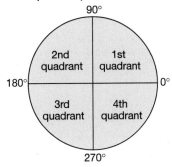

So we can say, for example, when θ has a value of 100° it is in the 2nd quadrant.

Similarly angles of 200° and 340° are in the 3rd and 4th quadrants respectively.

As P moves round the circumference of the circle, so x and y can be negative or positive depending on the value of θ: for example when θ is in the 2nd quadrant, x is negative, but y is positive.

This means that in the 2nd quadrant, $\sin \theta = \frac{y}{1}$ is positive, but $\cos \theta = \frac{x}{1}$ is negative (by convention we always assume that the radius is positive).

Check that you agree with this table.

Quadrant	1st	2nd	3rd	4th
sin θ	positive	positive	negative	negative
cos θ	positive	negative	negative	positive
tan θ	positive	negative	positive	negative

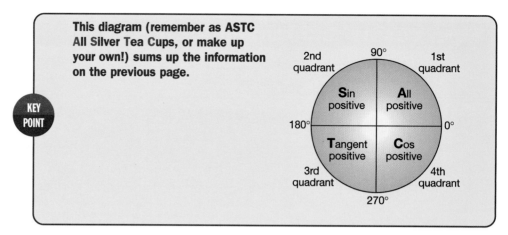

KEY POINT

This diagram (remember as ASTC All Silver Tea Cups, or make up your own!) sums up the information on the previous page.

The graphs of the three trigonometric ratios are shown below. You need to recognise them and distinguish between them. Note that all three ratios repeat every 360° (which corresponds to a complete revolution round the circle).

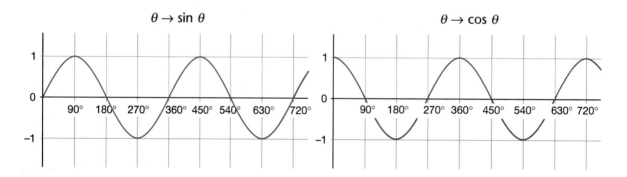

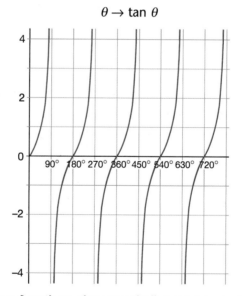

The sin and cos functions give very similar curves – both have maximum and minimum values of +1 and –1 respectively. But the tan function is of a different form and tends towards infinity around 90°, 270°, 450°, etc.

When a function like the sin function oscillates between –1 and +1 we say that it has an **amplitude of 1**. The graph of cos θ also has an amplitude of 1. The graph, for example, of 3cos θ, will have an amplitude of 3.

(Think back to plotting $y = f(x)$ and $y = 3f(x)$ in the plotting functions section, page 82.)

Example

Draw the graphs of $y = 3\cos x$ and $y = 2$.

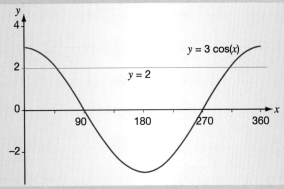

Use these graphs to solve the equation $1.5\cos x = 1$ for $0° \leq x \leq 360°$.

At the points where the two graphs intersect

$$3\cos x = 2$$

Dividing by 2 gives:

$$1.5\cos x = 1$$

The graphs intersect at approximately 50° and 310°.

Sine and cosine rules

The sine and cosine rules apply to all triangles – not just right-angled triangles. Although you will be given these in the exam, it is a good idea to remember them.

The sine rule

The sine rule is used to find the missing sides and angles in a triangle if (a) the length of one side and the sizes of two angles are known, or (b) the lengths of two sides and the size of the angle opposite one of those two sides are known.

KEY POINT

The sine rule:
$$\frac{a}{\sin A} = \frac{b}{\sin B} = \frac{c}{\sin C} \quad \text{or} \quad \frac{\sin A}{a} = \frac{\sin B}{b} = \frac{\sin C}{c}$$
Note how symmetrical each term is in each case.

Sketch and label the diagram first. Look carefully at what you know and what you need to find – make a rough but clear note of what you need to do. Questions like these are called multi-step, because you need to find something else before calculating the actual answer.

Examples

(a) Two surveyors are 25 m apart. One is North of a radio mast, and one is South. They each measure the angle of elevation of the top of the mast. One finds this to be 85°, the other 66°. Find the height of the mast to the nearest metre.

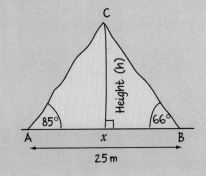

Use sine rule to find AC in triangle ABC.

In triangle ABC, using the sine rule:

$$\frac{b}{\sin B} = \frac{c}{\sin C}$$

Once AC is known you can use it in the right-angled triangle ACX to find h.

$$\text{or } \frac{AC}{\sin 66} = \frac{25}{\sin (180 - 85 - 66)}$$

$$\text{so } \frac{AC}{\sin 66} = \frac{25}{\sin 29}$$

$$\text{giving } AC = \frac{25 \times \sin 66}{\sin 29}$$

Don't work this out yet – leave it to the end.

In triangle AXC:

$$\sin 85 = \frac{\text{opposite}}{\text{hypotenuse}} = \frac{\text{height } (h)}{AC}$$

$$\text{so height } (h) = AC \times \sin 85$$

$$= \frac{25 \times \sin 66}{\sin 29} \times \sin 85$$

$$= 47 \text{ m to the nearest metre.}$$

(b) Find the two possible values of angle C in this triangle.

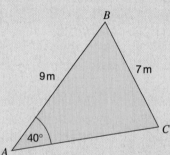

We know two sides and an angle opposite one of these sides so it is pretty certain that the sine rule is needed.

Using the sine rule in triangle ABC:

$$\frac{9}{\sin C} = \frac{7}{\sin 40}$$

By cross-multiplying.

$$9 \times \sin 40 = 7 \times \sin C$$

$$\text{so } \sin C = \frac{9 \times \sin 40}{7}$$

Using the inverse function key(s) gives this = 55.7° (3 s.f.).

$$= 0.8264 \ldots$$

$$\text{or } C = \sin^{-1} (0.8264\ldots) = 55.7°$$

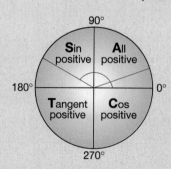

There are two possible angles having a positive sin as you can see from this diagram.

Check by finding the sin of 55.7° and 124.3° – they should be the same.

The other angle whose sin is $0.8264 \ldots$ is the angle $(180 - 55.7)° = 124.3°$.

So the two angles are 55.7 and 124.3°.

The cosine rule

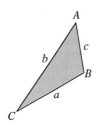

KEY POINT

Using the usual notation for angles and sides in a triangle – the cosine rule: $a^2 = b^2 + c^2 - 2bc \cos A$ or $b^2 = a^2 + c^2 - 2ac \cos B$ or $c^2 = a^2 + b^2 - 2ab \cos C$.

Look at how symmetrical the three versions are.

The cosine rule is useful when:

- you know two sides and the angle between them

- you only know the lengths of all three sides and need to find an angle.

Examples

(a) A plane takes off from an airport, flies for 50 km in a straight line, then turns through 20° and carries on for another 73 km.

How far, to the nearest km, is the plane from the airport as the crow flies?

> A sketch is crucial here.

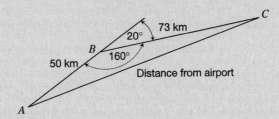

Using the cosine rule in ABC, with AC as the distance from the airport, gives:

> Make sure you can see how the cosine rule has been applied here.

$AC^2 = AB^2 + BC^2 - 2 \times AB \times BC \times \cos 160$

| Take care over negative numbers, e.g. cos 160° is negative

So $AC^2 = 50^2 + 73^2 - 2 \times 50 \times 73 \times \cos 160$

$= 14\,688.75\ldots$

giving $AC = 121$ km to the nearest km.

(b) A triangle has sides of length 12, 7 and 8 cm.

Find the largest angle in the triangle.

> The largest side in any triangle is always opposite the largest angle.

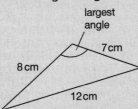

> We are given all three sides and no angles so the cosine rule should be used.

Using the cosine rule $a^2 = b^2 + c^2 - 2bc \cos A$, which can be rearranged to give:

$$\cos A = \frac{b^2 + c^2 - a^2}{2bc}$$

The largest angle is opposite the 12 cm side,

giving the size of the largest angle as:

$$\cos (\text{largest angle}) = \frac{7^2 + 8^2 - 12^2}{2 \times 7 \times 8} = -0.276\,78\ldots$$

so the angle is 106° to the nearest whole number.

1 Find the missing angles and sides for each of these shapes.

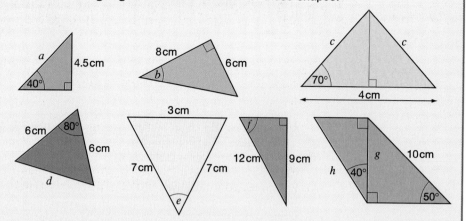

2 A square-based pyramid has base $ABCD$ and vertex V vertically above the middle of the base. $AB = 5$ cm and $VC = 10$ cm.

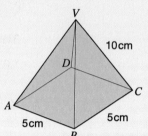

Find:
(a) DB
(b) the height of the pyramid
(c) the angle between VB and the base $ABCD$.

3 This is a sketch of a level triangular piece of land. $AB = 75$ m, $AC = 60$ m and angle BAC is an acute angle.

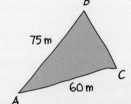

The piece of land has an area of 2000 m².
(a) Calculate angle BAC.
(b) Calculate BC to the nearest metre.

4 Solve these trigonometry equations. Find all the solutions between 0° and 360°.
(a) $\sin x = 0.2$ (b) $2\cos x = 1.5$ (c) $4\tan x = 17$
5 The diagonals of a rhombus are 8 cm and 4 cm in length.
Find the angles of the rhombus.
6 In the triangle ABC, angle $BAC = 23°$, angle $ABC = 68°$ and $AB = 5.6$ cm.
Find the lengths of the sides AC and CB.
7 In the triangle ABC, $AB = 17$ cm, $BC = 23$ cm and angle $ABC = 43°$.
Find AC.

7 15.7 cm
6 $AC = 5.19$ cm, $CB = 2.02$ cm
5 The angles are 127° and 53° (nearest degree).
4 (a) 11.5°, 168.5° (b) 41.4°, 318.6° (c) 76.8°, 256.8°
3 62.7°, 71 m
2 (a) 7.07 cm (b) 9.35 cm (c) 69.3°
(h) 10.0 cm
1 (a) 7.00 cm (b) 36° (c) 5.84 cm (d) 7.71 cm (e) 24.7° (f) 48.6° (g) 7.66 cm

3.4 Congruent triangles

LEARNING SUMMARY

After studying this section, you will be able to:
● *identify congruent triangles*
● *understand and use the conditions for two triangles to be congruent*

Congruent triangles

KEY POINT

Two triangles are congruent if they are identical in every way except for their position; one can be turned into the other by rotation, translation or reflection.

Two triangles are congruent if any of these sets of conditions are satisfied. (Equal sides and angles are indicated.)

● All three sides are the same **(SSS)**.

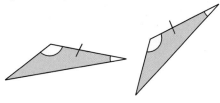

● Two equal pairs of sides and the angles they form are equal **(SAS)**.

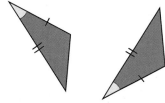

● A pair of equal sides and the angles at either end match **(ASA)**.

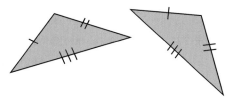

● Each triangle is right-angled and one side of one triangle is equal to the hypotenuse and a side of the other triangle: right angle, hypotenuse, side **(RHS)**.

Showing that two triangles are congruent by definition means that corresponding pairs of sides and angles are equal in size.

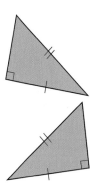

Example

$ABCD$ is a parallelogram.

Prove that opposite angles are equal (e.g. angle ABC = angle ADC).

Joining BD allows use to be made of the parallel line results.

In triangles ABD and DBC:

angle CBD = angle BDA (alternate angles)
angle ABD = angle BDC (alternate angles)
side BD is common to both triangles

So BDA and BDC are congruent (ASA).

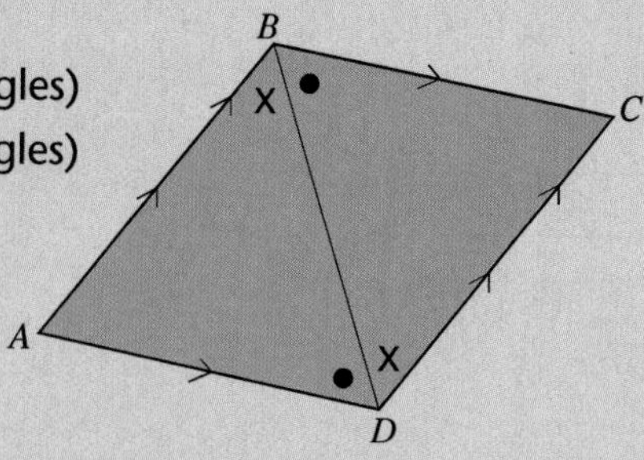

Angle ABC and angle ADC are corresponding pairs of angles, so angle ABC = angle ADC.

We have not laid down any conditions other than $ABCD$ is a parallelogram, so this result is true for all parallelograms.

PROGRESS TEST

1 In each of these, state if the triangles ABC, XYZ are congruent. If they are, state which congruency condition applies (SAS, SSS, etc.):
(a) $AB = XY$, $AC = XZ$ and angle BAC = angle YXZ
(b) Angle BAC = angle YXZ, angle ABC = angle XYZ, angle ACB = angle YZX
(c) $AC = XZ$, $BC = YZ$, angle BAC = angle YXZ
(d) $AC = XZ$, $AB = XY$, angle BAC = angle XYZ
(e) Angle CBA = angle YZX = 90°, $CA = XY$, $BC = ZX$

2 Use congruent triangles to prove that diagonals of a rectangle are equal.

3 $ABCDE$ is a regular polygon. X and Y are on BC and CD respectively, and $CX = CY$.
Show that triangles ABX and triangle EDY are congruent.
How do you know that $AX = EY$?

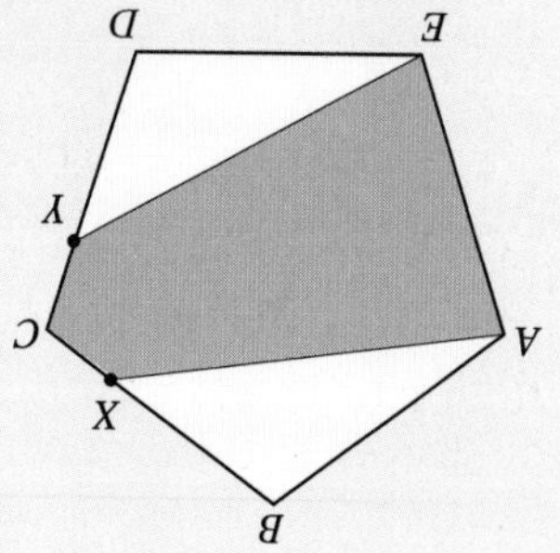

3 $AB = ED$ (sides of regular pentagon)
Angle ABX = angle EDY (angles of regular polygon)
$DY = BX$ ($DY = DC - CY$, $BX = BC - CX$, but $CX = CY$ (given) and $BC = CD$ (regular polygon)
So triangles BAX and EDY are congruent (SAS) ($AX = EY$ as they are a corresponding pair of sides in two congruent triangles)

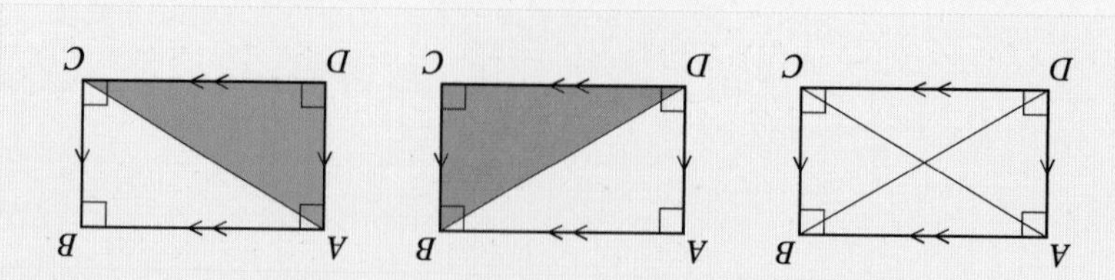

2 Triangles BCD and ADC are congruent (SAS). DB and AC are corresponding sides so $DB = AC$

1 (a) Yes, SAS (b) No (c) No (d) No (e) Yes, RHS

3.5 Circles

LEARNING SUMMARY

After studying this section, you will be able to:
- *find angles in cyclic quadrilaterals*
- *use the angle and tangent properties of circles including*
    angles at the centre and circumference of a circle
    angles between tangents and chords
    angles in the alternate segment

Parts of a circle

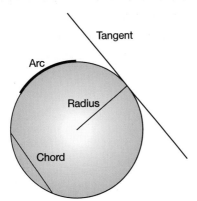

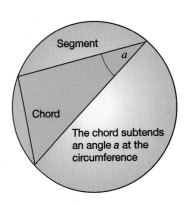

Cyclic quadrilaterals

A cyclic quadrilateral has vertices that all lie on a circle.

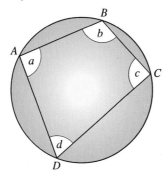

KEY POINT

Opposite angles in a cyclic quadrilateral add up to 180°:

$$a + c = 180° \text{ and } b + d = 180°$$

Angle facts involving the circle

You need to remember the following facts about angles in circles. (In each diagram, angles with dots in are equal.)

You may be asked to use any of these facts to prove or justify results.
Give reasons like 'opposite angles in a cyclic quadrilateral' or 'angles in same segment are equal'.

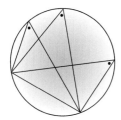

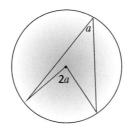

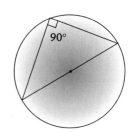

Angles in the same **segment** are equal or **angles subtended** by the same arc/**chord** are equal.

The angle at the centre is twice the angle subtended at the circumference.

The angle at the semicircle is 90°.

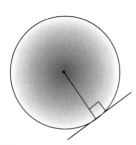

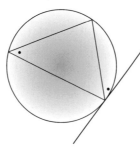

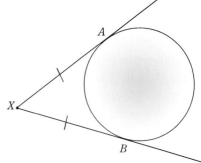

The angle between a tangent and the radius where it touches the circle is 90°.

The angle between a tangent and any chord where it touches the circle is equal to the angle in the alternate segment.

Tangents from a point (e.g. *X*) are equal in length (*XA* = *XB*).

Examples

The straight line *ABC* is a tangent to the circle centre *O*.

BOE is a diameter of the circle. Angle *FEO* = 50° and angle *DBC* = 30°. Find the sizes of the lettered angles.

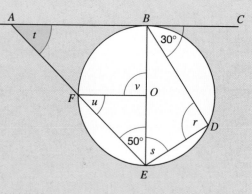

r = 90° (angle subtended by a diameter)

s = 30° (angle in alternate segment)

t = 40° (*ABE* is a right-angled triangle – *AC* is a tangent and angle *AEO* = 50°)

u = 50° (*FOE* is an isosceles triangle – radii form two sides)

v = 100° (angle *FOE* = 180 – 50 – 50, angle *FOB* = 180 – angle *FOE*)

PROGRESS TEST

1 *O* is the centre of this circle and *AB* is parallel to *CD*.
Find the angles labelled *x* and *y*.

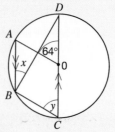

2 *O* is the centre, *BC* and *BA* are tangents to the circle from *D*.
Calculate the values of angles *a* and *b*.

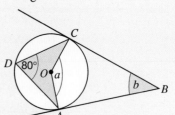

2 *a* = 160°, *b* = 20°

1 *x* = 32°, *y* = 58°

114

3.6 Proof

After studying this section, you will be able to:
● understand and read simple geometric proofs
● prove some geometrical facts met with previously

Using proof in geometry

KEY POINT A proof is a careful, logical argument using facts that are known to be true to show that something else is true.

One way of writing out mathematical proofs is to use two columns; in the left-hand column is a list of statements and in the right-hand column is the reason for the corresponding statement. The reason may be one of the givens for the proof, or a well-known theorem (a mathematical statement that has been already been proved), or one of the earlier statements from the left-hand column.

These examples are all proofs of facts that you should know.

Examples
Prove these using only the parallel line facts and the fact that a straight line is 180°.

(a) **Sum of the interior angles of a triangle is 180°** ($a + b + c = 180°$).

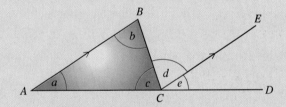

$a = e$ (corresponding angles)
$b = d$ (alternate angles)
so $a + b = e + d$
therefore $(a + b) + c = e + d + c$
but $e + d + c = 180°$ (ACD is a straight line)
so $a + b + c = 180°$

(b) **Exterior angle of a triangle is equal to the sum of the two interior opposite angles,** or using the above diagram, angle BCD = angle BAC + angle ABC.

$a = e$ (corresponding angles)
$b = d$ (alternate angles)
so $a + b = (e + d)$
or angle BAC + angle ABC = angle BCD – which is the required result.

(c) **The sum of the interior angles of an *n*-sided polygon is $(n - 2) \times 180°$**

The interior angles of a triangle sum to 180° from result (a).

Looking at these polygons it can be seen that an *n*-sided polygon may be split up into $(n - 2)$ triangles.

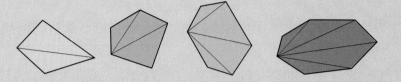

So the interior angle sum of an *n*-sided polygon will be $(n - 2) \times 180°$.

Congruent triangles are also very useful in helping to prove results:

(d) **The diagonals of a rhombus bisect each other at right angles.**

Let the diagonals cross at *E*. By definition all the sides of a rhombus are equal, with opposite sides parallel.

In triangle *ABE* and triangle *DEC*:
  angle *EAB* = angle *ECD* (alternate angles)
  angle *ABE* = angle *EDC* (alternate angles)
    AB = *DC* (given property of rhombus)

So the triangles *ABE* and *DEC* are congruent (ASA).

This means that *DE* = *EB* (corresponding sides of congruent triangles).

Therefore *DB* is bisected by *AE*.

In triangle *DAE* and triangle *BAE*:
  DE = *ED* (proved above)
  AE = *AE* (common to both triangles)
  DA = *AB* (given property of rhombus)

So triangles *DAE* and *BAE* are congruent (SSS):

angle *DEA* = angle *AEB* (corresponding sides of congruent triangles)

But *DEB* is a straight line so angle *DEA* and angle *AEB* are right angles.

So *DB* is bisected by *AE* and *AE* is at right angles to *DB*.

It can also be proved, using exactly the same method, that *AC* is bisected by *BD*.

(e) **If a line through the centre of a circle bisects a chord that is not a diameter, it is also perpendicular to the chord.**

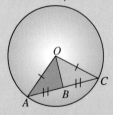

In triangles *OAB* and *OBC*:
  OA = *OC* (radii of circle)
  OB = *OB* (common to both)
  AB = *BC* (given, *B* bisects chord)

So triangles *OAB* and *OBC* are congruent (RHS).

The fact that radii are the same length in the same circle is a useful piece of information in some proofs involving circles.

Therefore angle OBA = angle OBC (corresponding angles in congruent triangles).

But ABC is a straight line, 180°.

Therefore angle PBA = angle OBC = 90°.

1 Prove that in any parallelogram, opposite sides are equal in length.
Use only the parallel line facts and congruent triangles.

2 Prove that the exterior angle e of a cyclic quadrilateral is equal to the interior opposite angle i.

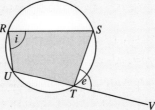

Use the fact that opposite angles of a cyclic quadrilateral sum to 180°.

3 This diagram shows a rectangle whose vertices all lie on a circle.

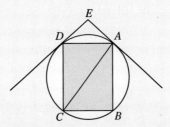

Tangents to the circle at D and A meet at E. Angle $DAE = 40°$.
(a) Explain why CA is a diameter of the circle.
(b) Calculate the value of (i) angle ADE and (ii) angle ACB – explain each step in your working.

4 Use the properties of isosceles triangles and this diagram to prove that the angle in a semicircle is a right angle.

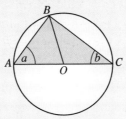

The centre of the circle is O.

5 Use the fact that the angle between the tangent and radius is 90° and congruent triangles to prove that $TA = TB$, where TA and TB are the tangents to the circle drawn from T.

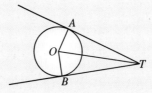

PROGRESS TEST

117

PROGRESS TEST

1 Triangles ABC and ACD are congruent (ASA) so $AB = CD$ (corresponding sides)
Similarly for AD and BD

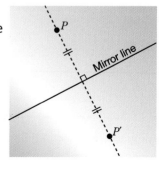

2 Angle URS + angle STU = 180° ($RSTU$ cyclic quad., opposite angles sum to 180°) but angle STU + angle STV = 180° (angles on straight line sum to 180°)
So angle URS = angle STV or $i = e$

3 (a) Angle CDA = 90° ($ABCD$ is a rectangle)
so CA must be a diameter (angle subtended by a diameter is 90°)
(b) (i) Angle DAE = angle ADE (DAE is isosceles triangle, $DE = EA$)
so angle ADE = 40°
(ii) Angle EAD = angle DCA (alternate segment)
Angle ACB = 90° − 40° = 50° ($ABCD$ rectangle, angles = 90°)

4 Triangles ABO and BOC are isosceles ($OA = OB = OC$ = radii of circle)
So angle ABC = (a + b)
In triangle ABC angle sum is $(a + b) + a + b = 2(a + b)$
but angle sum is 180°, $2(a + b) = 180$, so $(a + b)$ = 90°.

5 In triangles AOT and BOT
$OA = OB$ (radii of circle)
angle OAT = angle OBT = 90° (radius/tangent property)
$OT = OT$ (common to both triangles)
So triangles AOT and BOT are congruent (RHS)
Therefore $AT = BT$ (corresponding sides of congruent triangles)

3.7 Transformations

LEARNING SUMMARY

After studying this section, you will be able to:
- *recognise and describe fully the transformations: reflection, translation, rotation and enlargement*
- *identify and specify combinations of the above transformations*

Reflection

KEY POINT

To describe a reflection you need to state a single line, called the **axis of reflection** or **mirror line**.

Each point *P* maps onto its image *P′* so that the mirror line is the perpendicular bisector of *PP′*.

> Comparing the sense of object and image can be a useful test for whether or not a single reflection has taken place.

The image of an object after reflection is the same size as the object (shape is conserved), but the image has a reversed **sense** – the vertices of the image go in a clockwise order (*ABC*), in the image the order is anti-clockwise *A′B′C′*, or vice versa.

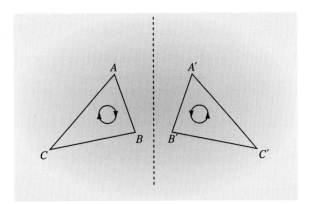

Translation

KEY POINT

A translation transforms an object by moving all points in that object by the same amount in the same direction.

The resulting image has the same size and sense as the object.

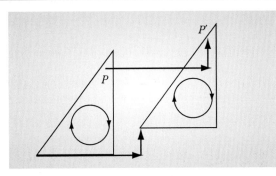

For the example above, the translation is described as '4 units to the right, then 1 unit up'.

You must give the across and up distances to specify the translation completely. It is usual to describe translations as a column vector, for the case here $\begin{pmatrix} 4 \\ 1 \end{pmatrix}$; a translation of 2 units to the left then 1 up would be described by $\begin{pmatrix} -2 \\ 1 \end{pmatrix}$.

Example

Translating the triangle ABC by $\begin{pmatrix} 5 \\ 3 \end{pmatrix}$ gives the triangle $A'B'C'$.

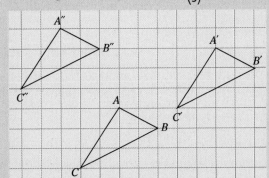

Translating $A'B'C'$ by $\begin{pmatrix} -8 \\ 1 \end{pmatrix}$ gives the triangle $A''B''C''$.

The translation $\begin{pmatrix} -3 \\ 4 \end{pmatrix}$ maps ABC onto $A''B''C''$, and combining the two translations above $\begin{pmatrix} 5 \\ 3 \end{pmatrix} + \begin{pmatrix} -8 \\ 1 \end{pmatrix} = \begin{pmatrix} -3 \\ 4 \end{pmatrix}$.

Rotation

 KEY POINT A rotation is completely described by giving the angle of rotation, the direction of rotation and the centre of rotation. Rotation preserves shape, size and sense, so an object and its image after rotation are congruent.

We take an anit-clockwise rotation to be positive and a clockwise rotation to be negative.

Examples

(a) Rotating triangle ABC through $-90°$ (clockwise) about $(-2, 1)$ gives the image $A'B'C'$.

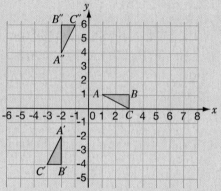

(b) Rotating $A'B'C'$ through $180°$ about $(-2, 1)$ gives $A''B''C''$.

(c) ABC can be mapped onto $A''B''C''$ by a rotation through $-270°$ about $(-2, 1)$.

You may also be asked to rotate a shape through any size of angle.

Example

Rotate triangle ABC through $50°$ about O.

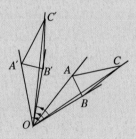

- From O draw lines to A, B and C.
- Draw lines from O making $50°$ to OA, OB and OC.
- Mark off $OA' = OA$, $OB' = OB$ and $OC' = OC$.
- Join A', B' and C' to complete the rotated image.

Alternatively measure OA and OB and draw OA' with $AOA' = 50°$ and use tracing paper to complete the image.

This drawing shows the rotation of the line AB by $60°$ anti-clockwise about O.

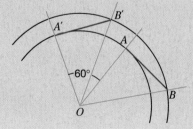

The image of AB after the rotation is $A'B'$.

When a figure is rotated about a centre of rotation:

- all the points in are turned through the same angle
- the points keep a constant distance from the centre of rotation.

This gives a method of finding the centre of rotation, given the object and its image.

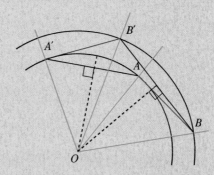

The line AA' is a chord to the circle whose centre is the centre of rotation. This lies on the perpendicular bisector of the chord, lines from two such chords will 'fix' the centre.

Enlargement

KEY POINT

An enlargement is completely described by giving the enlargement scale factor and the centre of enlargement. Enlargement preserves shape and sense, but not size. A reduction in size is an enlargement between −1 and 1, for example 0.5.

Example

O marks the centre of enlargement.
Triangle ABC is the object.

$OA' = 3 \times OA$, $OB' = 3 \times OB$ and $OC' = 3 \times OC$.

So $A'B'C'$ is an enlargement of ABC, centre of enlargement O and **scale factor 3**. (Note that OA and OA', etc. are in the **same direction**.)

$OA'' = 2 \times OA$, $OB'' = 2 \times OB$ and $OC'' = 2 \times OC$.

Note that OA and OA', etc. are in **opposite directions**.

So $A''B''C''$ is an enlargement of ABC, centre of enlargement O and **scale factor −2**.

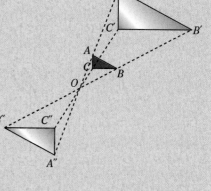

Combined transformations

The image generated by one transformation can be used as the object of a second transformation.

Examples

(a) A triangle A with vertices (1, 1), (1, 3) and (2, 3) is rotated through a quarter turn clockwise about (1, 1), then through a further quarter turn clockwise about (0, 0).

To what single transformation is this equivalent?

A rotated through a quarter turn clockwise about (1, 1) → B.

B rotated a quarter turn clockwise about (0, 0) → C.

A → C is a rotation of 180° about (1, 0).

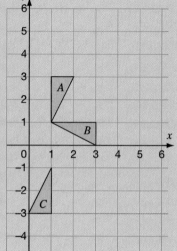

(b) When two transformations are combined, the order in which the transformations are applied can make a difference to the final image. If the initial image is the red triangle then:

Diagram 1: Rotation of –90° about (0, 0) (A) then reflection in the y-axis (B). Diagram 2: Reflection in the y-axis (A) followed by a rotation of –90° about (0, 0) (B).

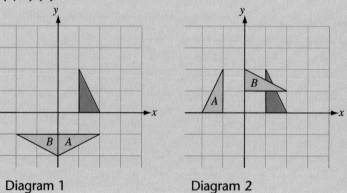

Diagram 1   Diagram 2

1   Draw a triangle ABC with vertices A(2, 4), B(8, 4) and C(8, 7).
    (a) Reflect ABC in the x-axis and draw its image $A'B'C'$.
    (b) Reflect $A'B'C'$ in the y-axis and draw the image $A''B''C''$.
    (c) What single transformation would map ABC onto $A''B''C''$?
2   The line AB has end-points A(2, 3) and B(3, 5).
    The line AB is mapped onto A'(3, 2) and B'(5, 3) by a single transformation.
    Draw the line and its image, describe the transformation fully.
3   Draw the triangle ABC with vertices A(2, 1), B(4, 2) and C(3, 3).
    The triangle is mapped onto $A'B'C'$ with vertices A'(7, 6), B'(11, 8) and C'(9, 10). Describe the transformation fully.
4   Draw the two points P(1, 4) and Q(1, 2). Join the two points to give the line PQ. Rotate the line PQ 30° anti-clockwise about the origin. Label the transformed line $P'Q'$. Now rotate $P'Q'$ 70° anti-clockwise about (1, 2) and label the transformed line $P''Q''$.
    What single rotation is equivalent to the combination of these two rotations?

3.8 Coordinates in three dimensions

LEARNING SUMMARY

After studying this section, you will be able to:
● *use and understand coordinates in three dimensions*

2D coordinates

Coordinates are numbers that describe the exact position of a point. A point in two dimensions (2D), such as on a page, can be located using a pair of axes at right angles, for example:

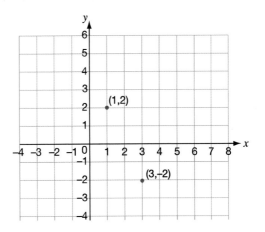

In 2D the distance of any point (x, y) from the origin is given by $\sqrt{x^2 + y^2}$ (using Pythagoras).

Similarly the distance between two points (x_1, y_1) and (x_2, y_2) is
$\sqrt{(x_1 - x_2)^2 + (y_1 - y_2)^2}$

Another useful result is the mid-point of the line whose end-points are (x_1, y_1) and (x_2, y_2), which is $\left(\dfrac{x_1 - x_2}{2}, \dfrac{y_1 - y_2}{2} \right)$.

3D coordinates

When you need to locate a point in three dimensions (3D) a third axis is used. This is the *z*-axis.

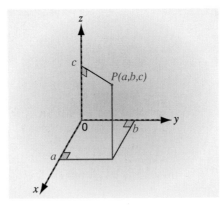

The position of a point is now specified by three numbers:
(*x*-coordinate, *y*-coordinate, *z*-coordinate).

Example
OABCDEFG is a cuboid, with *O*(0, 0, 0), *B*(0, 4, 2) and *F*(3, 4, 0).
Write down the coordinates of the other five vertices.

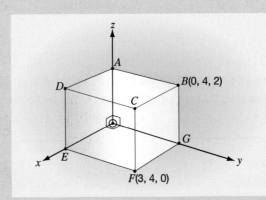

A(0, 0, 2)   *C*(3, 4, 2)   *D*(3, 0, 2)   *E*(3, 0, 0)   *G*(0, 4, 0)

In a similar way to 2D coordinates, the distance between two points (x_1, y_1, z_1) and (x_2, y_2, z_2) is $\sqrt{(x_1 - x_2)^2 + (y_1 - y_2)^2 + (z_1 - z_2)^2}$

To find distance of a point from the origin let the second point be (0, 0, 0).

KEY POINT   The coordinates of a point in 3D space are **x**, **y** and **z** in that order.

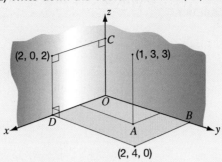
PROGRESS TEST

1 (a) Write down the coordinates of A, B, C and D.

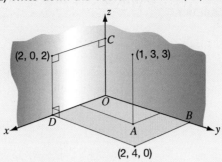

(b) Calculate the distance CD.
(c) Calculate the distance from the origin to (1, 3, 3).

2 A cube has sides of lengths 3 units. One vertex is fixed at (2, 1, 0).
   (a) How many different positions could the cube be placed in?
   (b) Write down the possible coordinates of the opposite vertex of the cube.

1 (a) A(1, 3, 0), B(0, 4, 0), C(0, 0, 2), D(2, 0, 0)
   (b) 2.83 units (3 s.f.)   (c) 4.36 units (3 s.f.)
2 (a) 8
   (b) (−1, −2, 3), (5, −2, 3), (−1, 4, 3), (5, 4, 3), (−1, −2, −3), (5, −2, −3), (−1, 4, −3), (5, 4, −3)

3.9 **Vectors**

LEARNING SUMMARY

After studying this section, you will be able to:
● **add and subtract vectors**
● **know what a scalar is**
● **solve geometrical problems with the aid of vectors**
● **solve problems involving forces with the aid of vectors**

Writing vectors

KEY POINT

A vector is a quantity that has both size and direction. Vectors are only equal if they have the same size (or magnitude) and direction. A vector can be thought of as a directed line segment. (A line goes on forever in both directions.) When we want to look at only a portion of a line, that has two ends, this is a line segment.

Vectors may be written in bold, e.g. **a**, or sometimes handwritten with a wavy line underneath like this: a̰ or labelled by their starting and end-points such as $\overrightarrow{AB}$.

They may also be written as a column of numbers such as $\binom{1}{2}$.

The four vectors here, shown as directed line segments, are all equal – they are clearly the same length and direction.

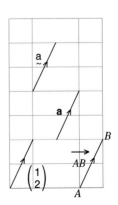

The vector –**u**, for example, is the same magnitude of vector **u**, but in the opposite direction:

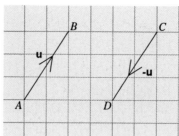

$\mathbf{u} = \begin{pmatrix} 2 \\ 3 \end{pmatrix}$ and $-\mathbf{u} = \begin{pmatrix} -2 \\ -3 \end{pmatrix}$ or $\vec{AB} = \vec{DC} = -\vec{CD}$

The **inverse** of a vector is a vector of equal magnitude but in the opposite direction.

So the inverse of $\vec{AB}$ is $-\vec{AB}$ or $\vec{BA}$.

Adding and subtracting vectors

> This is known as the triangle law.

To add two vectors means apply the first vector then apply the second vector. Two vectors can be added as shown here.

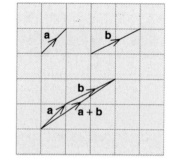

For this particular example:

$\mathbf{a} = \begin{pmatrix} 1 \\ 1 \end{pmatrix}$ and $\mathbf{b} = \begin{pmatrix} 2 \\ 1 \end{pmatrix}$

> The column vector and the drawing give the same answer.

So $\mathbf{a} + \mathbf{b} = \begin{pmatrix} 1 \\ 1 \end{pmatrix} + \begin{pmatrix} 2 \\ 1 \end{pmatrix} = \begin{pmatrix} 3 \\ 2 \end{pmatrix}$.

When two vectors are added or subtracted to produce a third vector, this vector is called the **resultant**.

Another way of looking at addition and subtraction of vectors is the **parallelogram law**, for the two vectors **m** and **n**:

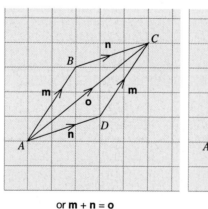

> The parallelogram law shows that going from *A* to *C* via *B* is the same as going from *A* to *C* via *D*

or **m** + **n** = **o**   or **m** – **n** = **p**

$\vec{AB} + \vec{BC} = \vec{AC}$ is the same as $\vec{DA} + \vec{AB} = \vec{DB}$ is the same as
$\vec{AD} + \vec{DC} = \vec{AC}$ $-\vec{AD} + \vec{AB} = \vec{DB}$

Subtracting a vector is the same as adding its inverse: **a** – **b** is the same as **a** + (–**b**).

Multiplying vectors

Scalars are quantities which have magnitude but not direction. A vector may be multiplied by a scalar, for example:

$\mathbf{a} = \begin{pmatrix} 2 \\ 1 \end{pmatrix}$, so $3\mathbf{a} = 3\begin{pmatrix} 2 \\ 1 \end{pmatrix} = \begin{pmatrix} 3 \times 2 \\ 3 \times 1 \end{pmatrix} = \begin{pmatrix} 6 \\ 3 \end{pmatrix}$

or, using line segments,

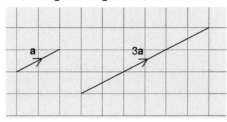

Vectors can be used to prove geometric properties and relationships. Most errors arise when students forget that, for example, $\vec{AB}$ is not equal to $\vec{BA}$, it is equal to $-\vec{BA}$; directions are crucial when working with vectors.

Examples

(a) In this parallelogram, express these vectors in terms of **p** and **q**.

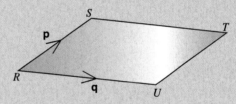

   (i) $\vec{SU}$ (ii) $\vec{US}$ (iii) $\vec{RT}$ (iv) $\vec{TR}$

(i) $\vec{SU} = -\mathbf{p} + \mathbf{q}$ (or $\mathbf{q} - \mathbf{p}$)   (ii) $\vec{US} = \mathbf{p} - \mathbf{q}$
(iii) $\vec{RT} = \mathbf{p} + \mathbf{q}$ (iv) $\vec{TR} = -\mathbf{p} - \mathbf{q}$ (or $-(\mathbf{p} + \mathbf{q})$)

(b) *ABCD* are the four vertices of a quadrilateral with $\vec{AB} = \vec{DC}$ and $\vec{AD} = \vec{BC} = \vec{BC}$

What can you deduce about the quadrilateral?

The quadrilateral is a parallelogram as opposite pairs of sides are parallel and equal in length (equal vectors for opposite sides so same magnitude (same length and direction (parallel)).

(c) *ABCD* is a quadrilateral. The mid-points of *AB, BC, CD* and *AD* are *W, X, Y* and *Z* respectively.

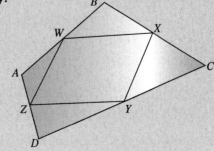

Show that $WXYZ$ is a parallelogram.

As W is the mid-point of AB and X is the mid-point of BC:

$\overrightarrow{WB} = \frac{1}{2}\overrightarrow{AB}$ and $\overrightarrow{BX} = \frac{1}{2}\overrightarrow{BC}$

so $\overrightarrow{WX} = \overrightarrow{WB} + \overrightarrow{BX} = \frac{1}{2}(\overrightarrow{AB} + \frac{1}{2}\overrightarrow{BC})$

Similarly

$\overrightarrow{ZD} = \frac{1}{2}\overrightarrow{AD}$ and $\overrightarrow{DY} = \frac{1}{2}\overrightarrow{DC}$

so $\overrightarrow{ZY} = \overrightarrow{ZD} + \overrightarrow{DY} = \frac{1}{2}(\overrightarrow{AD} + \overrightarrow{DC})$

But $\overrightarrow{AC} = \overrightarrow{AB} + \overrightarrow{BC} = 2\,(\overrightarrow{WX})$ and also $\overrightarrow{AC} = \overrightarrow{AD} + \overrightarrow{DC} = 2\,(\overrightarrow{ZY})$

Therefore $\overrightarrow{WX} = \overrightarrow{ZY}$, showing that these two sides are parallel and the same length.

Similarly for $\overrightarrow{WZ}$ and $\overrightarrow{XY}$, although proving $\overrightarrow{WX} = \overrightarrow{ZY}$ is all that is needed to prove $WXYZ$ is a parallelogram.

So $WXYD$ is a parallelogram.

Finding the resultant of two vectors is used to solve problems involving vector quantities such as force and velocity, but remember to give both magnitude and direction when asked for the resultant of two vectors.

Example

Find the resultant of two forces of magnitude 7 N and 13 N acting on a mass as shown in the diagram.

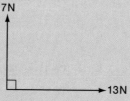

We need to add the two forces (vectors), using the parallelogram law:

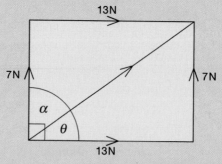

As the forces are acting at 90° the magnitude of the resultant is found using Pythagoras and is $\sqrt{218}$ N ($\sqrt{13^2 + 7^2}$), making an angle to the horizontal of θ where $\tan\theta = \frac{7}{13}$.

Therefore $\theta = 28.3°$ (to 3 s.f.)

In some cases you might be asked to use a scale diagram to find the resultant, rather than use trigonometry

1 In the diagram $\overrightarrow{OA}$ = **a**, $\overrightarrow{OB}$ = **b**, $\overrightarrow{OC}$ = $4\overrightarrow{OA}$ and $\overrightarrow{OD}$ = $k\,\overrightarrow{OB}$

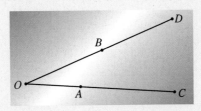

(a) Find in terms of **a**, **b** and k the vectors (i) $\overrightarrow{AB}$ (ii) $\overrightarrow{CD}$
(b) For what value of k is CD parallel to AB?

2 This diagram shows two forces pulling on an object.

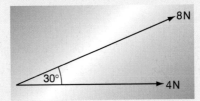

(a) Draw a vector diagram to scale to show the resultant force acting
   on the object.
(b) Hence or otherwise find the resultant force on the object.

3 ABCDEF is a regular hexagon.
$\overrightarrow{OA}$ = **a** and $\overrightarrow{OB}$ = **b**

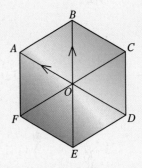

(a) Write down in terms of **a** and **b**:
   (i) $\overrightarrow{AB}$ (ii) $\overrightarrow{FC}$
(b) From your answers to (a) deduce one geometrical fact about AB and FC.

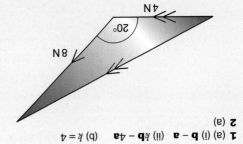

3.10 Constructions and loci

LEARNING SUMMARY

After studying this section, you will be able to:
- **understand and use the standard geometrical constructions**
- **construct a locus subject to given conditions**

Constructing triangle ABC from given information

When drawing constructions or loci, always leave in your construction lines particularly arcs of circles – otherwise there is no evidence that you used a construction, rather than just measured.

(a) **Given three sides:**

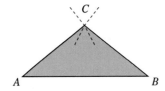

    (i)   Draw the line AB to given length.
    (ii)   Use compasses to construct arcs of radius equal to sides AC and BC.
    (iii)   Where the arcs intersect is where both sides fit – vertex C.

(b) **Given two sides and the included angle:**

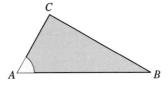

    (i)   Draw the line AB to the given length.
    (ii)   Measure and draw the angle at A.
    (iii)   Draw the line AC to the given length.
    (iv)   Join C to B.

(c) **Given two angles and the included side:**

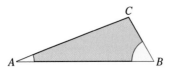

    (i)   Draw the line AB to the given length.
    (ii)   Measure and draw the given angles at A and B.
    (iii)   Draw the lines AC and BC.

The perpendicular bisector of a line

To bisect line AB:

(i) use a pair of compasses to draw arcs with the same radius from A and B,

(ii) join the points of intersection; the resulting line is the perpendicular bisector of AB.

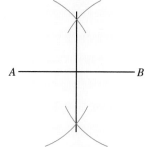

Why it works

> As well as knowing how to do the following constructions, you need to know *why* they work.

AX, AZ, BX and BZ are all equal radii.

Triangles AXZ and BXZ are congruent (SSS).

Angle AXY = angle BXY (corresponding angles).

So triangles AXY and BXY are congruent (SAS).

Therefore $AY = YB$.

Angle AYX = angle BYX (corresponding angles).

AYB is a straight line so angle AYX = angle BYX = 90°.

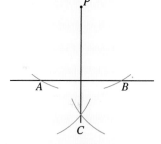

The perpendicular from a point to a line

To draw the perpendicular from P to a given line:

(i) from P draw arcs to cut the line at A and B

(ii) from A and B draw arcs with the same radius to intersect at C

(iii) join P to C: this is the perpendicular to the line AB.

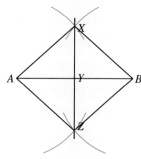

Why it works

$PA = PB$ (radii)

$CA = CB$ (radii)

Triangles PAC and PBC are congruent (SSS).

So angle APX = angle BPX (corresponding angles in congruent triangles).

Triangles PXA and PXB are congruent (SAS).

So angle PXA = angle PXB.

But AXB is a straight line.

So angle PXA = angle PXB = 90°.

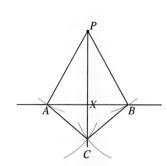

To bisect an angle

To bisect the angle at A:

(i) with centre A draw arcs to cut the lines at B and C

(ii) with the same radius draw arcs centre B and C to cut at D

(iii) join A to D.

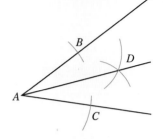

Why it works

$AB = AC$ (radii of same circle).

$BD = DC$ (radii of identical circles).

AD is common.

Triangles ABD and ACD are congruent (SSS).

Therefore angle BAD = angle CAD (corresponding angles of congruent triangles).

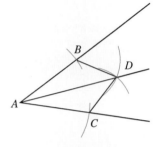

Loci

>
> **KEY POINT**
>
> **A locus is a set of points satisfying a particular condition; the condition is usually to do with their distance from certain fixed points, or the angles they make with other points and lines.**
>
> **For example, the locus of points at distance *r* from fixed point *C* is a circle radius *r* with centre at *C*. The plural of locus is loci.**

Three important loci are:

(1) The circle (see above): points equidistant from a fixed point, the centre.

(2) The perpendicular bisector: the locus of points that are equidistant from two fixed points (see page 131).

(3) The angle bisector: the locus of points that are equidistant from two fixed lines (see above).

Examples

(a) Two radio beacons, X and Y, are 40 km apart. X has a range of 25 km and Y has a range of 30 km. Make a sketch of the region in which both radio beacons can be received.

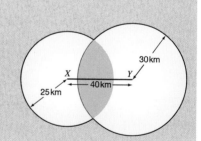

(b) A tree is to be planted in a garden. The garden is rectangular measuring 10 m by 8 m.

The tree must be equidistant from the walls along *AB* and *AD*. It must also be 3 m from the wall along *DC*.

Sketch the situation and show the point *P* where the tree must be planted.

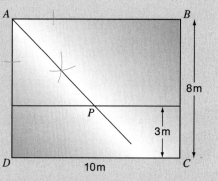

1 A distress call is heard by three boats *A*, *B* and *C*, shown in this sketch.

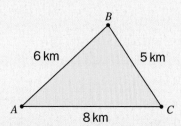

The ship in distress is nearer to ship *C* than to ship *A* and less than 3 km from ship *B*.

Draw an accurate plan of the position of the boats *A*, *B* and *C*.

Indicate all the possible positions of the ship in distress.

2 Draw a line 5 cm long.

Show the region in which all points are less than 4 cm from the line.

3 Construct this triangle accurately.

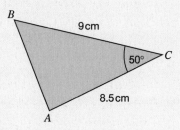

Measure the side *AB*.

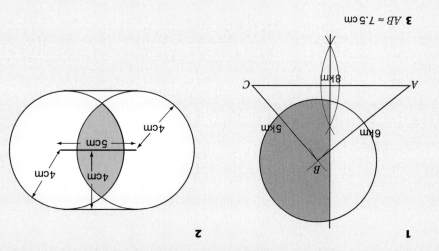

3 *AB* ≈ 7.5 cm

3.11 Dimensions, areas and volumes

LEARNING SUMMARY

After studying this section, you will be able to:
- distinguish between formulae for length, area and volume
- recognise whether or not a formula is consistent in terms of its units and why
- find areas and volumes of similar figures

Dimensions

Length is a measurement in **one dimension**: the dimension of, say, a perimeter is length (*L*).

Area is a measurement in **two dimensions**: the dimension of area is length × length (*L²*).

Volume is a measurement in **three dimensions**: the dimension of volume is length × length × length (*L³*).

KEY POINT — Any valid expression or equation must be **dimensionally consistent**.

Examples

(a) $2\,m^3 + 4\,m^3 = 6\,m^3$ is dimensionally consistent (all are L^3)

$2\,m^3 + 4\,m$ is not dimensionally consistent:

$2\,m^3$ has dimensions L^3 and $4\,m$ has dimensions L – adding a volume and a length does not make sense.

(b) The surface area (L^2) of a sphere is given by the equation $4\pi r^2$.

The dimensions of $4\pi r^2$ is (number) $\times L^2$ which is L^2 – it is dimensionally consistent.

(c) According to Pythagoras, h (a length, L) = $\sqrt{r^2 - s^2}$.

The dimensions of $\sqrt{r^2 - s^2}$ are $[L^2 - L^2]^{0.5}$ which is $[L^2]^{0.5}$ which is L.

The formula $h = \sqrt{r^2 - s^2}$, where h, r and s represent lengths, is dimensionally consistent.

Areas and volumes of similar figures

If two figures are similar then corresponding edges on those figures are in the same ratio.

In this example the ratio of the lengths of the sides of the two cubes is 2 : 3.

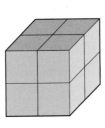
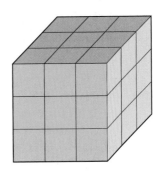

> **You may have worked out the areas as 2×2 and 3×3 already.**

The ratio of the areas of the faces on the two cubes is $4 : 9$ (check by counting). This is the same as $2^2 : 3^2$.

In the same way the ratio of the volumes of the two cubes is $2^3 : 3^3$ or $8 : 27$.

The above line of reasoning can be applied to any similar shapes or solids.

> **KEY POINT**
>
> Area scale factor is the **square** of the length scale factor.
>
> Volume scale factor is the **cube** of the length scale factor.
>
> For example: if a copy of a statue is to be made double the height of the original (i.e. length scale factor 2), the new statue will have 2^2 or 4 times the surface area and 2^3 or 8 times the volume (and mass).

Examples

> **Always check to make sure you are working in the same units.**

(a) The Eiffel Tower is 300 m high. A model of the Eiffel Tower is 30 cm high. The model has a base area of 100 cm².

What is the base area of the actual tower?

Length scale factor = 30 cm : 300 m which is $30 : 30\,000$ or $1 : 1000$

So the area scale factor will be $1 : 1\,000\,000$, giving the 'real' base area as

$100 \times 1\,000\,000$ cm²

or

> **Be careful with large powers of ten.**

$100 \times 1\,000\,000 \div 10\,000$ m²

= $10\,000$ or 10^4 m².

(b) Two spheres have volumes in the ratio $27 : 216$.

What is the ratio of their radii?

Volume scale factor is $27 : 216$

So length scale factor is $\sqrt[3]{27} : \sqrt[3]{216} = 3 : 6 = 1 : 2$.

Radii are in the ratio of $1 : 2$.

> **If length scale factor is $a : b$, volume scale factor is $a^3 : b^3$ (cube of length scale factor). So if volume scale factor is $r : s$, then length scale factor is $\sqrt[3]{r} : \sqrt[3]{s}$ (cube root of volume scale factor).**

(c) Two mathematically similar shapes have surface areas in the ratio $25 : 9$.

   (i)   What is the ratio of their volumes?

  (ii)   The larger solid has a volume of 500 cm³.

         What is the volume of the smaller?

   (i)   Area scale factor = $25 : 9$

         So length scale factor = $\sqrt{25} : \sqrt{9}$

                  = $5 : 3$

         Giving a volume scale factor of $5^3 : 3^3 = 125 : 27$.

  (ii)   So the volume of the smaller solid is $500 \times \dfrac{27}{125} = 108$ cm³.

PROGRESS TEST

1 In the following expressions r, a, d and h represent lengths.
For each expression state whether the expression represents a length, an area, a volume or none of these.

(a) $\frac{\pi}{4} \times d^2$ (b) $\frac{h\pi d^2}{12}$ (c) $2\pi r(r + h)$ (d) $\frac{\pi r a^3}{d}$

(e) $\pi a(r + h) + \pi(r^2 + h^2)$ (f) $radh$

2 Indicate which of these expressions are dimensionally inconsistent; a, b, c and d represent lengths.

(a) $a + bc$ (b) $ab + bc$ (c) $a^2 - cd$ (d) $\pi ab + ac$

3 Two similar cones have heights of 3 cm and 6 cm respectively.
The volume of the smaller cone is 30 cm³. Its surface area is 24 cm².
Find the (a) volume and (b) surface area of the larger cone.

4 Two similar cans of paint hold 1.5 litres and 2.5 litres of paint.
The height of the smaller can is 20 cm.
What is the height of the larger can?

1 (a) area   (b) volume   (c) area   (d) volume   (e) area   (f) none of these
2 Only (a) is inconsistent
3 (a) 240 cm³   (b) 96 cm²
4 23.7 cm (3 s.f.)

3.12 Volumes of prisms, cones, pyramids and spheres

LEARNING SUMMARY

After studying this section, you will be able to:
● *calculate the volumes of prisms, cones, pyramids and spheres*

Prism

You need to know and be able to use the formulae for the areas of the following shapes: parallelogram, rhombus and trapezium. You may also need them in working out the volumes of prisms.

A prism is any solid that can be cut into slices that are all the same shape.
The volume of a prism can be found by multiplying the area A of the regular cross-section (the end) by the length of the shape (L in these diagrams).

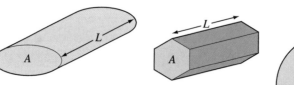

Prisms come is a great variety of shapes: for example, a cylinder or a flat puddle.

Cone

Volume $= \frac{1}{3} \times$ base area $\times$ perpendicular height

$$= \frac{1}{3}\pi r^2 h$$

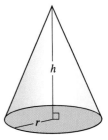

Pyramid

> **Note similarity with cone formula.**

> **The pyramid need not have a square base, or have its apex directly over the centre of the base.**

Volume $= \frac{1}{3} \times$ base area $\times$ perpendicular height

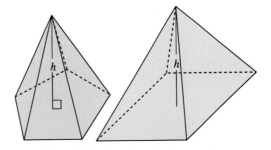

Sphere

Volume of a sphere is $\frac{4}{3}\pi r^3$
where r is the radius of the sphere.

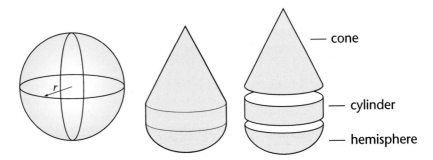

— cone

— cylinder

— hemisphere

When dealing with problems involving the volume of complicated solids, split them into simpler solids, then find the volume of these separately.

Example

The Great Pyramid is 155 m high. Its square base measures 229 m by 229 m.

What is the volume enclosed by the Great Pyramid?

Volume $= \frac{1}{3} \times$ base area $\times$ perpendicular height

$\qquad = \frac{1}{3} \times 229 \times 229 \times 155$

$\qquad = 2\,709\,452\ m^3$ (to the nearest whole number)

Example

A test tube consists of a cylinder with a hemisphere at one end.

The length of the cylinder is 20 cm, and its radius of cross-section is 1 cm.

Find the volume of the test tube.

Split the test tube up into a cylinder and a hemisphere:

Volume of the cylinder $= \pi \times 1 \times 20\ cm^2$

A hemisphere has half the volume of a sphere with the same radius.

Volume of hemisphere $= \frac{1}{2}(\frac{4}{3}\pi 1^3)\ cm^2$

So total volume $\qquad = \pi \times 1 \times 20 + \frac{1}{2}(\frac{4}{3}\pi 1^3)\ cm^2$

$\qquad\qquad\qquad\quad = 64.9\ cm^3$ (3 s.f.)

A frustum is what is left when a cone or pyramid has its top sliced off.

Many household items are frustums of particular solids.

Yoghurt pots are frustrums of cones.

Example

A large yoghurt pot in the shape of a frustum of a hollow cone, is made by removing a 42 cm part of the cone as shown.

Find the volume of the pot.

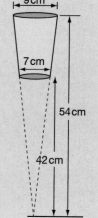

Volume of the complete cone

$\quad = \frac{1}{3} \times$ base area $\times$ perpendicular height

$\quad = \frac{1}{3} \times \pi \times 4.5^2 \times 54$

$\quad = 1145.1\ cm^3$ (1 d.p.)

Volume of cone removed $= \frac{1}{3} \times \pi \times 3.5^2 \times 42$

$= 538.8 \, \text{cm}^3$ (1 d.p.)

So volume of pot $= 1145.1 - 538.8$

$= 606 \, \text{cm}^3$ (to the nearest whole number)

PROGRESS TEST

1 Find the volume of a hemisphere of radius 30 cm.

2 A tunnel through a hill is 500 m long. It is in the shape of a semicircle of radius 4 m. Calculate the volume of earth removed when the tunnel was built.

3 A pyramid with a square base of side 8 cm has a volume of 320 cm³. What is the vertical height of the pyramid?

4 James drops a full 1 litre bottle of milk onto the kitchen floor. The milk forms a puddle 0.4 cm thick. What was the area of the puddle of milk?

5 This piece of jigsaw has a volume of 300 mm³. The puzzle is made out of wood 1.5 mm thick.

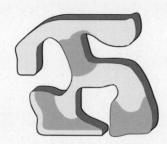

Calculate the area of one face of the piece.

6 A conical glass is 8.25 cm in diameter and 6 cm high.
   (a) What is its capacity in millilitres?
   (b) How many such glasses could be filled from a sherry bottle containing 0.75 litres?

1 56 500 cm³ (3 s.f.)
2 12 600 m³ (3 s.f.)
3 15 cm
4 2500 cm²
5 200 mm²
6 (a) 107 ml (b) 7 full glasses

3.13 Circles and spheres

LEARNING SUMMARY

After studying this section, you will be able to:
- *calculate the volumes and surface areas of spheres*
- *find the length of an arc*
- *recognise sectors and segments and calculate their areas*

The sphere

KEY POINT

The volume of a sphere $= \frac{4}{3}\pi r^3$

The surface area of a sphere $= 4\pi r^2$

(No need to muddle these up – just look at the dimensions.)

You have already met volumes of spheres in Section 3.12.

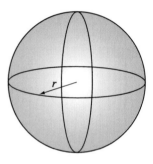

Examples

(a) Find the surface area of a hemisphere of radius 10 cm.

Area of curved surface $= \frac{1}{2}(4\pi r^2)$ which is $\frac{1}{2}(4\pi 10^2)$

$= 628.3 \text{ cm}^2$ (4 s.f.)

Area of circular base of hemisphere $= \pi r^2$ which is $\pi 10^2$

$= 314.2 \text{ cm}^2$ (4 s.f.)

> Don't forget the flat circular surface of the hemisphere.

So the surface area of hemisphere $= 628.3 + 314.2 \text{ cm}^2$

$= 943 \text{ cm}^2$ (to the nearest whole number)

(b) A metal sphere of radius 10 cm is melted down to make a cone of height 10 cm. What is the radius of the cone?

Volume of metal sphere $= \frac{4}{3}\pi r^3$

$= \frac{4}{3} \times \pi \times 10^3$

> It is usually a good idea to leave the actual calculation as late as possible.

Volume of cone $= \frac{1}{3}\pi r^2 h$

$= \frac{1}{3}\pi r^2 \times 10$

$\therefore \frac{1}{3}\pi r^2 \times 10 = \frac{4}{3} \times \pi \times 10^3$

$\frac{1}{\cancel{3}}\cancel{\pi} r^2 \times \cancel{10} = \frac{4}{\cancel{3}} \times \cancel{\pi} \times 10^{3\,2}$

so $r^2 = 4 \times 10^2$

$\therefore r = 2 \times 10 = 20 \text{ cm}$

Arcs, sectors and segments

An **arc** is part of the circumference of a circle.

The arc that subtends the larger angle at the centre is called the **major arc**.

The arc that subtends the smaller angle at the centre is called the **minor arc**.

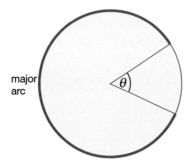

The length of an arc is directly proportional to the angle at the centre and the radius of the circle:

$$\text{arc length} = \frac{\theta}{360} \times \pi d \text{ or } \frac{\theta}{360} \times 2\pi r$$

A **sector** of a circle is an area bounded by two radii and an arc.

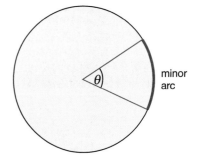

In the same way as the arc length is a fraction of the circumference of a circle, so the area of a sector is a fraction of the area of the circle.

The area of the minor sector shown is $\frac{\theta}{360} \times \pi r^2$.

A **segment** of a circle is the area bounded by a chord and an arc.

> Careful you don't confuse sector and segment – especially when you think of a sector of a piece of apple pie for example.

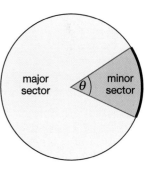

To work out the area of a minor segment, first work out the area of the sector and then subtract the area of the triangle *ABO*.

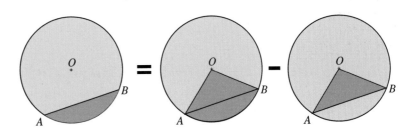

Similarly for the area of a major segment:

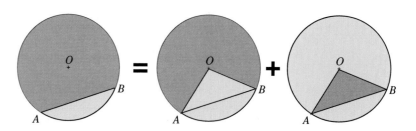

Example

A cylindrical oil storage tank has a diameter of 2 m.

The tank is 5 m long.

A volume indicator floats on the surface of the oil.

The indicator makes an angle of 50° to the vertical.

Calculate the volume of oil in the tank.

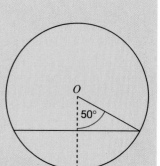

> **Making a sketch of the situation, and lettering angles, etc., helps clarify your thinking.**

We need to find the area of shaded segment *ABY*.

To do this, first we need to find the area of sector *OAYB*:

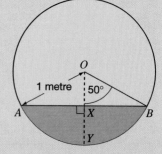

$$= \frac{\theta}{360} \times \pi r^2$$

which is $\frac{100}{360} \times 1^2 = \frac{100\pi}{360}$

> **See page 104 for $A = \frac{1}{2}ab \sin C$.**

Area of triangle $OAB = \frac{1}{2}ab \sin C$

which is $\frac{1}{2} \sin 100$.

So the area of shaded segment $= \frac{100\pi}{360} - \frac{1}{2}\sin 100 = 0.3803 \text{ m}^2$ (4 s.f.)

$\therefore$ Volume of oil = area of cross-section × length

$$= 0.3803 \times 5$$
$$= 1.90 \text{ m}^3 \text{ (3 s.f.)}$$

PROGRESS TEST

1 Calculate the area and length of arc of this sector.

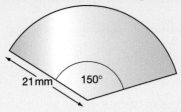

21 mm   150°

2 Calculate the area of the minor segment in this diagram.

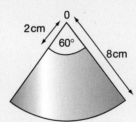

A   4 cm   120°   *O*   *B*

3 This logo, the shaded region, is part of a circle centre *O* and radius 8 m.

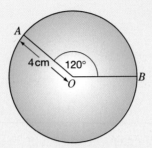

O   2 cm   60°   8 cm

Calculate
(a) the area of the logo
(b) the perimeter of the logo.
(Hint: don't forget the straight edges.)

4 Find the angle that subtends an arc length equal to the radius of the circle.

5 What is the surface area of a sphere whose volume is 200 cm³?

5 165.4 cm² to 2 d.p.

4 Angle $= \frac{360}{2\pi}$, 57.3° (3 s.f.)

3 (a) 31.4 cm² (3 s.f.)   (b) $(6 + 6 + 2\pi \times 2 \times \frac{60}{360} + 2\pi \times 8 \times \frac{60}{360}) = 22.5$ cm (3 s.f.)

2 16.8 cm² (to 3 s.f.)

1 (a) 577 mm² (nearest whole number)   (b) 55 mm (nearest whole number)

Sample GCSE questions

1 Calculate the volume of this block of metal. **[3]**

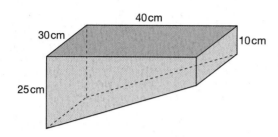

$\text{Area of trapezium shaped cross-section} = \frac{1}{2} \times (25 + 10) \times 40$ ✔

$= 700 \text{ cm}^2$ ✔

$\text{Volume of prism} = 700 \times 30$

$= 21\,000 \text{ cm}^3$ ✔

2

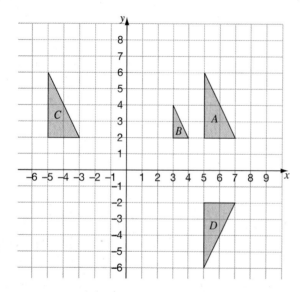

Describe fully the following single transformation that:

(a)　maps A onto D

(b)　maps A onto B

(c)　maps A onto C **[6]**

(a)　*Reflection in the x-axis (or y = 0)* ✔

(b)　*Enlargement, scale factor 0.5,* ✔✔
　　centre of enlargement (1, 2) ✔

(c)　*Translation,* $\begin{pmatrix} -10 \\ 0 \end{pmatrix}$ ✔✔

Always give a full description of any transformation.

In case you thought A to C was a reflection – check the sense of the vertices (the way they go round) – they are the same – so not an odd number of reflections.

Sample GCSE questions

3 (a) Write down the coordinates of the points A, B and C. **[3]**

(b) Calculate the distance of P from the origin. **[2]**

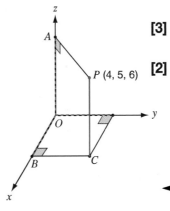

(c) Write down the coordinates of a point half way along the straight line from the origin to P. **[1]**

(a) $A(0, 0, 6)$ $B(4, 0, 0)$ $C(4, 5, 0)$ ✔✔✔

(b) $PO^2 = \sqrt{4^2 + 5^2 + 6^2}$ ✔
    $= 8.77$ (to 3 s.f.)   ✔

(c) $(2, 2.5, 3)$ ✔

4 Two straight lines AB and CD bisect each other at E.

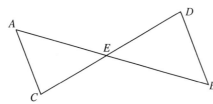

Prove that $AC = BD$. **[4]**

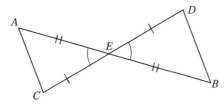

Using triangles AEC and BED

Angle AEC = angle DEB (vertically opposite angles)   ✔

AE = EB and CE = ED (given that AB and CB bisect each other)   ✔

Triangles AEC and BED are congruent (SAS)   ✔

∴ AC = BD (corresponding sides in two congruent triangles)   ✔

Sample GCSE questions

5 The base of a pyramid is a horizontal square, *ABCD*, with side 10 cm.

All the sloping edges are 15 cm long meeting at *E*.

The mid-point of the base is at *O*, and *E* is vertically above *O*.

Calculate the volume of the pyramid. Give the units of your answer. **[6]**

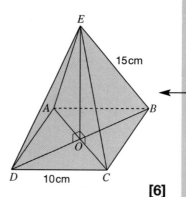

> *Do not assume facts about shapes from the diagram – it will nearly always say 'not to scale' – this should warn you. In particular beware of these: two lines appearing parallel, angles which look like right-angles, sides of triangles which appear equal, polygons that look as if they are regular.*

To find volume, you need to work out the vertical height EO.

In triangle DCB

Using Pythagoras:

$$DB^2 = BC^2 + CD^2$$
$$= 10^2 + 10^2$$

so $DB = \sqrt{200}$

But $DO = OB$

so $DO = \dfrac{\sqrt{200}}{2}$

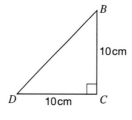

✔

✔

> *A rough sketch of the triangle you are dealing with can help.*

In triangle DOE

$$ED^2 = DO^2 + OE^2$$
$$so\ 15^2 = \frac{200}{4} + OE^2$$

giving $OE^2 = 15^2 - 50 = 175$

so $OE = \sqrt{175}$

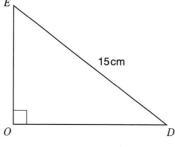

✔

✔

✔

> *Don't work out square roots until as late as possible – they might cancel out.*

Volume of pyramid $= \dfrac{1}{3} \times$ *base area* $\times$ *perpendicular height* ✔

which gives $\dfrac{1}{3} \times 10 \times 10 \times \sqrt{175} = 441\ cm^3$ *(to 3 s.f.)* ✔

> *Remember the units – usually 1 mark for correct units.*

Sample GCSE questions

6 An isosceles right-angled triangle has its two equal sides 1 unit long.

Use the triangle to find the value of:
(a) tan 45° **[1]**
(b) cos 45° **[2]**
Leave your answer in surd form.

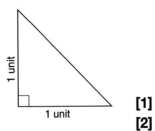

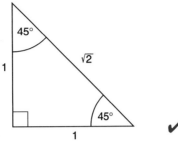

$tan45 = \frac{1}{1} = 1$ ✔

Using Pythagoras find the hypotenuse $= \sqrt{2}$ ✔

$Cos\ 45 = \frac{1}{\sqrt{2}}$ ✔

7

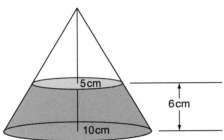

The diagram above shows a frustum of a cone. The radius of the base is 10 cm. The radius of the top is 5 cm. The height of the frustum is 6 cm.

Calculate the volume of the frustum. Leave your answer as a multiple of π. **[4]**

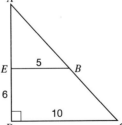

Triangles ABE and ACD are similar so $EB:DC = AE:AD$
∴ $5:10 = AE:(AE+6)$
So $AE = 6$ and so $AD = 12$ ✔
Volume of whole cone $= \frac{1}{3}\pi \times 10^2 \times 12$
$= 400\pi$ ✔
Volume of top small cone $= \frac{1}{3}\pi \times 5^2 \times 6$ ✔
∴ Volume of frustum $= 400\pi - 50\pi = 350\pi$ ✔

Sample GCSE questions

8 (a) Make an accurate copy of this triangle using ruler and compasses. **[2]**

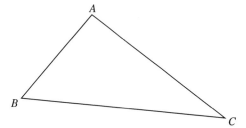

(b) Using ruler and compasses, construct:
  (i) the bisector of angle ABC **[2]**
  (ii) the perpendicular bisector of AB. **[2]**

(c) X is any point which satisfies the following conditions:
  (i) it is inside triangle ABC
  (ii) it is equidistant from BA and BC
  (iii) it is nearer to A than to B.

By clearly shading on your triangle, show clearly all the possible positions for X. **[2]**

After drawing a construction, do not rub anything out. The lines and arcs are part of your answer, and the examiner will want to see them.

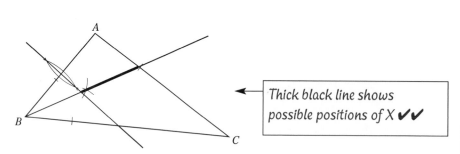

Copy of triangle, vertices within about 2 mm ✔✔

Bisector of angle ABC with arcs clearly visible ✔✔

Perpendicular bisector of AB with arc clearly visible ✔✔

Thick black line shows possible positions of X ✔✔

9 Air traffic control at O picks up two aircraft, A and B, on bearings $040°$ and $200°$ respectively. A is 40 km from traffic control; B is 30 km from traffic control.

How far are the two aircraft apart? **[3]**

Sample GCSE questions

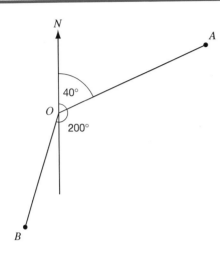
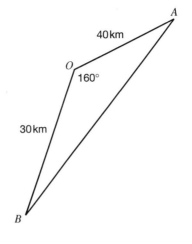

A rough sketch helps you to see what is needed.

Two sides and an included angle are known – very good chance that cosine rule is needed.

Using cosine rule:

$AB^2 = OA^2 + OB^2 - 2 \times OA \times OB \times \cos 160$
$AB^2 = 40^2 + 30^2 - 2 \times 40 \times 30 \times \cos 160$
$\qquad = 4755 \text{ (to 4 s.f.)}$
$\therefore AB = \sqrt{4755} = 69.0 \text{ km (to 3 s.f.)}$

✔

✔
✔

Always state what you intend to do and what needs working out – never keep all your working on your calculator – what's not seen won't get credit!

10 The diagram below shows the vectors **a**, **b** and **c**.

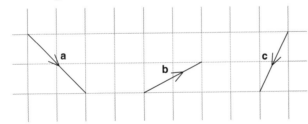

On some squared paper draw *and* label vectors to represent:

(a) $\overrightarrow{AB} = \mathbf{a} + \mathbf{c}$ [1]

(b) $\overrightarrow{CD} = \mathbf{a} + 2\mathbf{b}$ [1]

(c) $\overrightarrow{EF} = \mathbf{a} - \mathbf{b}$ [1]

(d) $\overrightarrow{GH} = \frac{1}{2}\mathbf{a} + \mathbf{b} - \mathbf{c}$ [1]

Remember direction matters with vectors $\overrightarrow{AB}$ and $\overrightarrow{BA}$ are not equal.

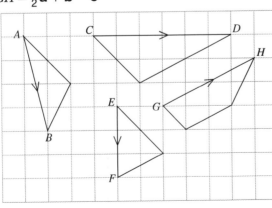

✔✔✔✔

Exam practice questions

1 $ABCD$ is an isosceles trapezium (i.e. $AD = BC$).
       angle $ACD = x°$.

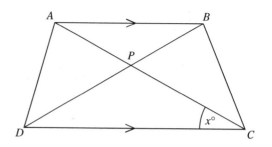

    (a)   Find the size of:
         (i)   angle ABD
         (ii)   angle APD.   **[3]**
    (b)   Prove that $ABCD$ is a cyclic quadrilateral.   **[4]**

2   In the diagram, $OACB$ is a parallelogram, XY is parallel to OB.
       X is the mid-point of OA and N is the mid-point of XY.
       $\overrightarrow{OA} = \mathbf{a}$ and $\overrightarrow{OB} = \mathbf{b}$

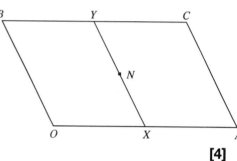

    (a)   Write in terms of $\mathbf{a}$ and $\mathbf{b}$:
         $\overrightarrow{XN}, \overrightarrow{ON}, \overrightarrow{AN}, \overrightarrow{NB}$ **[4]**
    (b)   Deduce two facts about the points A, N, B.   **[2]**

3   This is a torus (like a doughnut in shape). The radius of the inner circle is r and the
       distance from the centre of the ring to the centre of the hole in the torus is R.

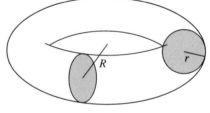

    One of these formulae gives the volume of a torus.

    $(2\pi Rr)^2$ $4\pi^2R$ $2\pi^2Rr^2$ $4\pi^2Rr$ $4\pi^2(r + R^2)$

    Which is the correct formula?

    Give a reason for your answer.   **[2]**

4 $ABCD$ is a regular tetrahedron of side 10 cm.

    AO is the perpendicular height of the tetrahedron.

    DO meets BC at M, where $BM = MC$, and $MO = \frac{1}{3}MD$.

    Calculate the volume of the tetrahedron.

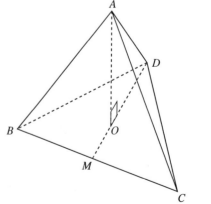

                                                                             [6]

Exam practice questions

5 (a) Sketch the curve $y = \sin x$ for values of x from $x = 0°$ to $360°$.
 Label the axes clearly. **[2]**

 (b) Write down the coordinates of the points of intersection with the x-axis. **[1]**
 Draw the line $y = 0.75$ on your sketch graph.
 The x-coordinates of the points where the line intersects the curves form the solution to an equation in x.

 (c) Write down this equation. **[1]**
 (d) Solve the equation. **[2]**

6 This diagram shows three quadrilaterals A, B and C.

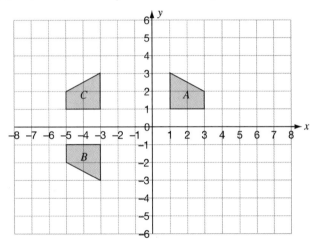

 (a) Describe fully the transformation that maps A onto B. **[3]**
 (b) Describe fully the transformation that maps B onto C. **[1]**
 (c) Describe fully the single transformation that maps A onto C. **[2]**

7 $ABCD$ are four points on the circumference of a circle, centre O.
 BOD is a straight line. Angle $BDC = 40°$ and angle $ABD = 65°$.

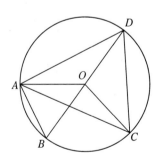

 Calculate the following angles, giving reasons for your answers:
 (a) Angle BAC **[1]**
 (b) Angle DAC **[2]**
 (c) Angle AOD **[2]**
 (d) Angle OAC **[3]**

8 James cleans his teeth twice a day. Each time he brushes his teeth he squeezes out a 'cylinder' of toothpaste 8 mm in diameter and 2 cm long.
 The toothpaste he buys comes in 100 ml tubes.
 How long should a tube of toothpaste last James? **[4]**

Exam practice questions

9   Hollow steel balls are used in the game of boules.

Each ball has an outside diameter of 80 mm.

The thickness of the steel is 5 mm.

Calculate the volume of steel needed to make one ball.   **[4]**

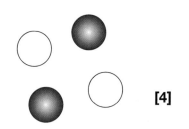

10   Amy is running an orienteering course. Here is a sketch showing the start and the first two check points.

The first checkpoint is 900 m due North from the start.

The second checkpoint is on a bearing 250° from the start and on a bearing of 200° from the first check point.

Calculate the straight line distance from the first to the second checkpoint.

Give your answer to a sensible accuracy.

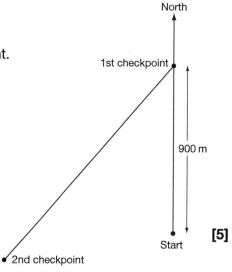

[5]

Overview

Topic	Section	Studied in class	Revised	Practice questions
4.1 Identification and selection	Populations, samples and bias			
	Sampling			
	Primary and secondary data			
4.2 Collecting data	Questionnaires and surveys			
	Two-way tables			
4.3 Processing and representing data	Stem and leaf diagrams			
	Scatter diagrams			
	Mean, mode and median of grouped data			
	Time series			
	Moving averages			
	Cumulative frequency			
	Quartiles			
	Box and whisker plots			
	Histograms			
4.4 Interpreting and examining data	Comparing data and drawing conclusions			
4.5 Probability	Independent events			
	Mutually exclusive events			
	Tree diagrams			

4.1 Identification and selection

LEARNING SUMMARY

After studying this section, you will be able to:
- *recognise bias*
- *understand sampling and use different sampling techniques*
- *understand the differences between primary and secondary data*

Bias

A **population** is an entire set of objects, observations or scores that have something in common. For example, a population might be all the females between the ages of 15 and 18 in the UK, all Year 10 students in a school or the population of Wales and so on.

It is not possible, for example, to ask all the people in Wales a particular question, so instead a **sample** of people in Wales is asked the question. It is important that this sample is a fair representation of all the people in Wales. Anything that distorts data so that it does not give a fair or representative picture of the population is called bias. Only if the sample is **unbiased** can conclusions drawn from it be applied to the whole population.

The actual questions asked in surveys can introduce bias. Imagine a TV company carrying out a survey into a new game show on national TV. Two ways in which bias could arise are:
- Asking the **leading question**, 'Have you seen that great new game show that everyone who is cool is watching?'
- Asking five people – the **sample size** is too small.

A biased sample would also be collected if:
- The questionnaire was only given to people in a shopping mall on a Monday morning. This would miss out people who were working at the time and possibly students and people at school as well.

Sampling

There are several methods used to collect a sample from a population.

Always show or state the type of sampling you are using and why you chose it.

1 Systematic sampling

An example of this method is the selection of a 1% sample by going through the population or list picking every hundredth item or individual. However, this produces a representative sample only if the population is arranged in a random way or as long as the list does not contain any hidden order, such as being in alphabetical or some other order.

2 Attribute sampling

In this method the sample uses some attribute that has no connection with what is being investigated. Choosing a sample to investigate TV watching habits on the basis of their birthday being the first of the month is an example of attribute sampling.

3 Stratified sampling

A stratified sample is obtained by taking samples from each stratum or sub-group of a population. A stratum or sub-group could be gender, year group or age group. When a population has several strata, the proportion of each stratum in the sample should be the same as in the population. For example, if the population contained twice as many vegetarians as non-vegetarians, then the sample should contain twice as many vegetarians as non-vegetarians.

4 Random sampling

With random sampling each member is chosen entirely by chance and each member of the population has an equal chance of being included in the sample. If you wanted to select a sample of 100 in a school with 2000 students, you might put all their names in a draw and then pull 100 names out.

5 Quota sampling

This is a method of sampling used in opinion polls and market research. Interviewers are each given a quota of subjects of specified type to interview, for example, an interviewer might be asked to select 20 adult men and 20 adult women, 10 teenage girls and 10 teenage boys so that they could interview them about their television viewing.

6 Cluster sampling

The entire population is divided into groups, or clusters, and a random sample of these clusters is selected. All observations in the selected clusters are included in the sample. It is used when the researcher cannot get a complete list of the members of a population they wish to study but can get a complete list of groups or 'clusters' of the population. In order to investigate the use of pesticides by farmers in England, a cluster sample could be taken using the different counties in England as clusters. A sample of these counties (clusters) would then be chosen at random, so all farmers in those counties selected would be included in the sample. It is easier and cheaper to visit several farmers in the same county than it is to travel to each farm in a random sample to observe the use of pesticides.

7 Stratified random sampling

A stratified random sample is obtained by:
- separating the population into appropriate categories or strata, e.g. age
- finding out what proportion of the population is in each stratum
- selecting a sample from each stratum in proportion to the stratum size.

This can be done by random sampling (see above), so the technique is called stratified random sampling. An example of using stratified sampling would be a survey about cloning human cells. If you wanted to reflect the diversity of the population, you should try to include people of various minority groups, such as ethnic or religious groups, based on their proportion in the total population. A stratified survey could be said to be more representative of the population than a survey based on random or systematic sampling.

Example

A detailed survey is to be carried out among students in Years 11, 12 and 13 about major changes in the school organisation. Only 20 students can be interviewed. There are 150 students in Year 11, 193 in Year 12 and 107 in Year 13. The sample is to be stratified, based on year group size. The sample size from each year group must be proportional to the stratum size.

There is a total of $150 + 193 + 107 = 450$ students.

> **Round to the nearest student.**

> **Round 8.577 down to keep total at 20.**

Year	Fraction of students	Number of students in the sample of 50 to the nearest whole number
11	$\frac{150}{450}$	$\frac{150}{450} \times 20 = 7$
12	$\frac{193}{450}$	$\frac{193}{450} \times 20 = 8$
13	$\frac{107}{450}$	$\frac{107}{450} \times 20 = 5$

So the investigator needs to select **at random** 7 students from Year 11, 8 from Year 12 and 5 from Year 11. (Just to check: $7 + 8 + 5 = 20$.)

Primary and secondary data

Primary data is collected **first hand** by the person who is going to analyse and use it. With a **secondary data** source the information is **second hand**. Primary data can be relied on because you know where it came from and what was done to it. Secondary data is data that is available from an external source, such as books, newspapers, the internet, TV, radio and stories and recollections told by people. Secondary data is cheaper and easier to acquire than primary data, but must be treated with caution. A ship's list of people who boarded the ship before it sailed is primary data, but an index (created by someone using the original list) is secondary data.

Example

Which of the following are primary and which secondary data?

(a) Looking at football handbooks to see how many famous footballers were born during May.
(b) Measuring the outside temperature every day at noon.
(c) Looking at the local free newspaper for the temperature last Monday.
(d) Finding out information about holidays on the internet.
(e) Comparing supermarket prices by noting the prices of the same items in supermarkets.
(f) Comparing supermarket prices by looking in the local free newspaper.

Primary: (b), (e)
Secondary: (a), (c), (d), (f)

1 Why might these methods give biased samples?

(a) A market researcher is investigating the effectiveness of a mobile phone display in a shop. He counts the number of people stopping to look at the display from 1 pm to 2 pm on a Monday and Friday.

(b) A council wants to try to encourage greater use of the sports centre. They carry out a survey of people using the sports centre on a Monday morning.

(c) A publisher wants to find out about people's book-buying habits. She interviews a sample of people at several large city libraries.

2 For each of the following write down which sampling method is being used to carry out a survey of students in Years 12 and 13 in a sixth form college.

(a) Listing all the students in Years 12 and 13 in alphabetical order, and then choosing the first and every fifth student after that.

(b) Numbering all the tutor groups from 1 to 10. Writing the numbers 1 to 10 on slips of paper and putting them into a bag. One slip is taken out without looking. The number on it gives the tutor group to be surveyed.

(c) Choosing the first ten boys and ten girls from each of the year groups who are in the dinner queue.

(d) Listing all the students in Year 12 and Year 13, giving each a number and using random numbers to select the sample.

(e) There are 150 students in Year 12 and 165 students in Year 13. The sample contains 10 students from Year 12 and 13 from Year 11 chosen at random.

3 This table shows the numbers of boys and girls in Year 7 and Year 8 of a school.

	Year 7	Year 8
Boys	115	128
Girls	110	115

The School Council wants to find out their views about possible changes to the timetable. They take a stratified random sample of 40 pupils from Year 7 and Year 8.

(a) Calculate the number of students to be sampled in Year 8.

(b) Calculate the number of girls to be sampled in Year 8.

4.2 Collecting data

 LEARNING SUMMARY

After studying this section, you will be able to:
- **construct and use questionnaires**
- **make and use two-way tables**

Questionnaires and surveys

There are three main ways to gather primary data: by using a **questionnaire**, direct **observation** or **experiment**. There are some general rules about using each of them.

A **questionnaire** should:

- give sufficient choices to cover all the possible answers
- be clear and easy to understand by the subject (the person who is answering it)
- have questions that are short and capable of being answered simply – yes/no responses are the best as these are easy to administer and to analyse
- be made up of questions relevant to the survey
- respect the subjects' confidentiality
- not have any biased or leading questions
- not take too long to complete.

When using **observation** consider the following:
- Are you actually answering the question asked?
- Does the timing of the observation matter?
- Did you collect the data for long enough?

When doing an **experiment** you need to answer these questions:
- Does the experiment really test the concept or hypothesis?
- Are there sufficient accurate results to draw a valid conclusion?

Questionnaire design

To sum up, a good questionnaire is easy to complete and easy to analyse. Here are some examples of both good and poor questions. Always have at the back of your mind the possibility of bias creeping into your questions.

Questionnaire

Question	Comment
Do you not find it unreasonable to not make too much noise after 10 pm?	The language is very hard to understand
What is your favourite pizza topping?	A reasonable question, but the responses will be far too varied to analyse.
Which is your favourite pizza topping? Tick one box. ☐ Cheese ☐ Tuna ☐ Mushroom ☐ Chicken ☐ Spicy beef ☐ None of these	A satisfactory question, easy to complete and easy to analyse.
How old are you? ☐ 11–16 ☐ 16–18 ☐ Over 21	The age groups overlap and there is no group for those aged 19 and 20.
Who is your favourite group? ☐ Zeek or ☐ Abca?	At first sight a good question, but it is open to abuse, e.g. if Zeek got more positive responses then 'Zeek is UK's most popular group'.
All sensible people like dogs. Do you like dogs?	This is biased. It implies you aren't sensible if you don't like dogs.

Once a survey has been performed (or even before it has begun) you need to think about how to present your results. There is a great deal of choice: bar chart, scatter graph, histogram, pie chart, etc.

Two-way tables

Two-way tables are used to show two sets of information.

Example

A PE teacher conducted a survey into how many students were able to swim.

	Can swim	Cannot swim	Total
Year 9	121	60	**181**
Year 10	178	14	**192**
Year 11	192	5	**197**
Total	**491**	**79**	**570**

> The grand total in the bottom right cell can be arrived at by adding either across or down the other totals – a useful check.

The table shows quite clearly that the number of non-swimmers changes with year group, and that the numbers in each year group are slowly increasing.

A common question involving two-way tables is to ask you to fill in the blank cells.

Example

This table shows some of the results of a survey which asked adults of different ages if they considered themselves as liberal, moderate or conservative in their choice of films. Complete the table.

	Under 30	30–49	50 and above	Total
Liberal		119	88	290
Moderate	140		284	704
Conservative	73	161		448
Total			586	

The missing values are shown in red.

	Under 30	30–49	50 and above	Total
Liberal	**83**	119	88	290
Moderate	140	**280**	284	704
Conservative	73	161	**214**	448
Total	**296**	**560**	586	**1442**

1 Comment on each of these questionnaire questions:
  (a) Is it not true that you don't eat enough vegetables?
      ☐ yes ☐ no
  (b) How many books do you read?
      ☐ a lot ☐ a fair number ☐ not as many as I should
  (c) What is your favourite film?
  (d) A healthy normal persons eats at least five vegetables a day.
      How many vegetables do you eat?
2 In a class of 32 students, there were 8 girls who played volley ball and 5 boys who did not. There were 15 girls in the class. How many boys played volley ball? [Use a two-way table to help you.]
3 Copy and complete this two-way table showing the numbers, in millions, of boys and girls of different ages in the UK.

Age	Boys	Girls	Total
Under 1		0.3	0.6
1–4	1.4		
5–14	3.9	1.3	
Total	5.6	5.3	10.9

PROGRESS TEST

1 (a) Almost impossible to understand, doesn't say what enough vegetables are.
(b) Although good use of answer boxes, the responses are not precise enough.
(c) This will invite a very large number of responses and will be almost impossible to analyse.
(d) Biased question and it will also produce a large number of different answers. Tick boxes would have been useful.

2 Putting the information into a two-way table:

	Boys	Girls	Total
Plays volley ball	12	8	20
Does not play volley ball	5	7	12
Total	17	15	32

12 boys played volley ball

3

Age	Boys	Girls	Total
Under 1	0.3	0.3	0.6
1–4	1.4	3.7	5.1
5–14	3.9	1.3	5.2
Total	5.6	5.3	10.9

4.3 *Processing and representing data*

After studying this section, you will be able to:
- *construct and interpret stem and leaf diagrams*
- *plot and use scatter graphs*
- *calculate the mean, median and mode for grouped data*
- *recognise and interpret time series*
- *calculate moving averages*
- *draw and interpret cumulative frequency curves, box plots and histograms*

Stem and leaf diagrams

A stem and leaf diagram provides a quick way to gain a visual summary of your data, and allows you to detect any interesting features in their distribution.

Example

Here are the marks gained by 29 students in a test.

26	40	42	17	24	29	31	34	35	45	46	49	26
59	27	36	39	42	47	49	56	62	24	25	25	21
12	14	14

For these figures the first digit represents the stem

Steps for constructing a stem and leaf diagram

1 Select one or more of the leading digits to be the stem values; the remaining digits become the leaves.

2 List possible stem values in a column.

```
1|
2|
3|
4|
5|
6|
```

3 Record the leaf for every observation beside the corresponding stem value:

```
1|7244
2|649674551
3|14569
4|02569279
5|96
6|2
```

4 Rewrite the leaves in numerical order

1|2447
2|144556679
3|14569
4|02256799
5|69
6|2

> For 1|2 the 1 is the stem (representing 10) and the leaf is 2.

5 Indicate in a key what units are used for the stems and leaves.

> For example write by the side of the table 1|2 represents 12

From the completed stem and leaf table it is easy to read off certain information:
- the modal group, the one with the highest frequency, is the 20–29 group.
- there are 29 results, so the median result is the $(29 + 1) \div 2 = 15$th result in order, starting with 12, which is 34.

> **KEY POINT** The stem is the first part of the number. The leaf has the end digits of the readings as in the example above and the one below.

Example

The following table gives the mass, in grams, of 20 small pebbles in a sample.

| 3.4 | 4.3 | 3.9 | 4.8 | 4.2 | 4.9 | 5.2 | 5.4 | 5.5 | 5.0 |
| 5.3 | 5.4 | 6.6 | 6.2 | 6.6 | 7.1 | 8.3 | 8.2 | 6.3 | 6.7 |

(a) Make a stem and leaf table to show the masses.

(b) Use your table to find (i) the modal group and (ii) the median mass. (See page 165.)

(a) The completed stem and leaf table is:

3|4 9
4|2 3 8 9
5|0 2 3 4 4 5
6|2 3 6 6 7
7|1
8|2 3   Key 6|8 represents 6.8 g

(b) (i) The modal group is 5.0 – 5.9

(ii) The median (half way between the tenth and eleventh values) is 5.4.

Scatter diagrams

A scatter diagram or scatter graph is used to plot two sets of data to see whether a connection or **correlation** can be established between them. It is a useful form of graph to use when you are trying to prove a **hypothesis** such as 'faster cars cost more' or 'the more hours of homework a student does, the better their GCSE results will be'.

Example

This table shows the amount in £s per household spent per week in 1980, on alcohol and tobacco, in 11 UK regions.

Alcohol (£)	6.47	6.13	6.19	4.89	5.63	4.52	5.89	4.79	5.27	6.08	4.02
Tobacco (£)	4.03	3.76	3.77	3.34	3.47	2.92	3.20	2.71	3.53	4.51	4.56

This data is plotted on a scatter graph, and it suggests that there is a relationship between spending on alcohol and spending on tobacco: the more spent on alcohol the more spent on tobacco. The mean amount spent on each is calculated: alcohol £5.44, tobacco £3.62 (both to the nearest p), which gives the 'mean point', which is plotted. A straight line is drawn through this mean point and as close as possible to the other points. This line is called the **line of best fit by eye**.

> **This scatter graph was drawn using a spreadsheet – useful if you want to use scatter graphs in your coursework. Most spreadsheet programs will also draw a line of best fit for you!**

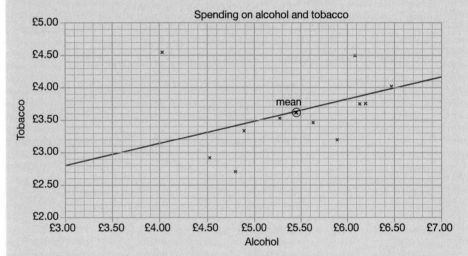

Using the line of best fit it is possible to make an estimate of money spent on alcohol for a particular amount spent on tobacco. For example, reading from the line of best fit, a household spending £5 a week on alcohol might be expected to spend about £3.50 on tobacco.

Correlation

Correlation is a measurement of how strong a relationship is between two sets of data. There are different kinds of correlation:

- **Negative correlation**. In this example the value of a car **decreases** as its age **increases**.

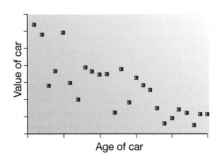

- **Positive correlation**. In this example the cost of a rail journey **increases** as the length of journey **increases**.

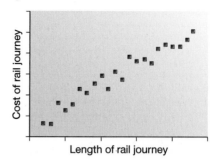

- **No correlation** or **zero correlation**. In this example there is no correlation between maths mark and height.

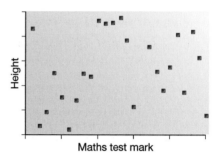

When the points on a scatter graph are not closely clustered in a straight line, we can have **weak positive correlation** and **weak negative correlation**:

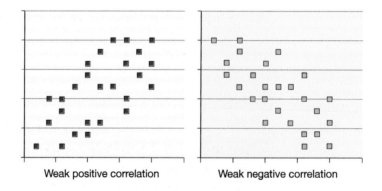

Mean, mode and median of grouped data

This example shows how to calculate the mean, mode and median from grouped data.

Example

Amy is completing a project about sentence length in newspapers. She keeps a record of the number of words in sentences chosen at random from a newspaper. Here are her results for one newspaper.

Σ is the Greek letter sigma, and is mathematical shorthand. It means 'sum' so, for example, Σz means the sum of all the zs, Σf means the sum of all the frequencies, etc.

Number of words	Frequency (f)	Mid-value (x)	$f \times x$
1–5	27	3	$27 \times 3 = 81$
6–10	25	8	$25 \times 8 = 200$
11–15	21	13	$21 \times 13 = 273$
16–20	10	18	$10 \times 18 = 180$
21–25	11	23	$11 \times 23 = 253$
26–30	4	28	$4 \times 28 = 112$
31–35	1	33	$1 \times 33 = 33$
Total	**99** ($= \Sigma f$)		**1132** ($= \Sigma fx$)

Modal class

Because the data is grouped, we use the term **modal class** rather than mode. For this example the modal class is 1–5 words – this group has the highest frequency.

Median

As the data is grouped we have to estimate the value of the 'middle value' (median). There are 99 values so the middle value is the 50th value, which occurs in the 6–10 group. This group is 5 numbers wide: 6, 7, 8, 9, 10. The frequency is 25 so the group has 25 values in it. The 50th value is the 23rd of these 25 values because $50 = 27 + 23$ (the 27th value is at the end of the 1–5 group) so an estimate would be $\frac{23}{25}$ of the way through this group. So the

median $= 5 + \frac{23}{25} \times 5$, giving 10 to the nearest whole number.

Mean

It is not possible to find the exact value of the mean when working with grouped data. It can be estimated by using the mid-value of each group as representing the whole group. We know that there are 16 words with a length between one and five letters (1, 2, 3, 4 or 5), so we use an estimated value of 3 – the mid-value. These estimated values for each group are shown in the table as the x-values.

Multiply each x-value by the frequency, f, of the group, as shown in the last column of the table.

Add up the values to fx to give Σfx.

Divide by the total frequency, Σf.

Remember to divide by Σf **not** the number of rows in the table.

So the estimated mean $= \dfrac{\Sigma fx}{\Sigma f} = \dfrac{1132}{99} = 11.4343$ or 11 to the nearest whole number.

Time series

A time series is made up of numerical data recorded at intervals of time, usually equally spaced (e.g. every minute or every hour). These points are joined in order, giving a line graph. The diagrams below show examples of two types of time series.

Graph A shows random fluctuations.

Graph B shows seasonal variations (to be expected with soft drink sales), with an upward movement or trend, shown in the drawn trend line.

A

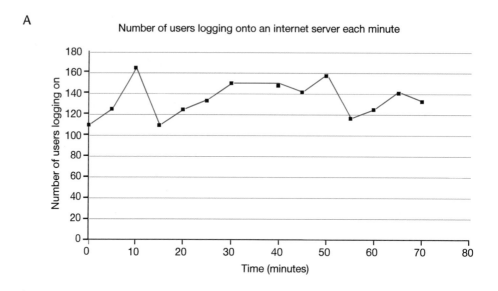

B

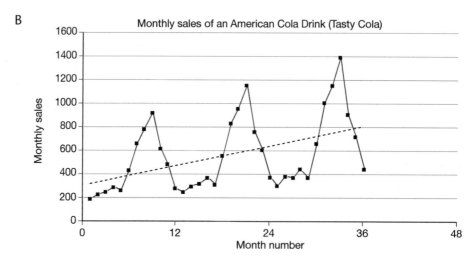

Example

The diagram below shows the number of cars produced in the UK in each quarter (Q1, Q2, etc.) from 2001 to 2005.

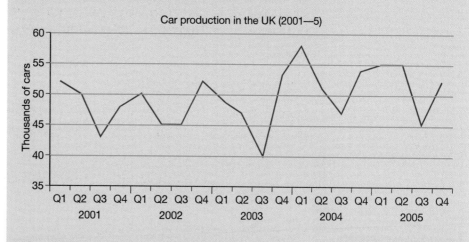

(a) How many cars were produced in 2001?

(b) What was the percentage change in car production between 2001 and 2005?

(a) Total for 2001 = Q1 + Q2 + Q3 + Q4 = 52 + 50 + 43 + 48

= 193 (thousand)

(b) 2001 = 193 000   2005 = 207 000

so percentage increase = $\dfrac{(207 - 193) \times 100}{193}$ = 7.25% (to 2 d.p.).

Moving averages

A moving average gives a clearer indication of the trend of a set of data. It smoothes out, for example, seasonal or monthly differences, as seen in the car production data above. A four-point moving average uses four data items in each calculation, a three-point moving average uses three and so on.

> Note how the 'oldest' number moves out and the 'newest' one moves into the list. You can use this to simplify the calculation of moving averages.

Example

Find the three-point moving averages for the following data:

9, 10, 5, 16, 15, 8, 13

For the first three values, 9, 10, 5, the average (mean) is (9 + 10 + 5) ÷ 3 = 8.

The next three values are 10, 5, 16 **moving along** one value), with an average of (10 + 5 + 16) ÷ 3 = 10.3.

The next three values are 5, 16, 15 with average (5 + 16 + 15) ÷ 3 = 12, and so on until you reach the last value, which for this data is 13.

The complete list of three-point moving averages is:

(9 + 10 + 5) ÷ 3 = 8   (10 + 5 + 16) ÷ 3 = 10.3   (5 + 16 + 15) ÷ 3 = 12
(16 + 15 + 8) ÷ 3 = 13 (15 + 8 + 13) ÷ 3 = 12

Moving averages can help you to draw a trend line on a time-series graph.

Example

Here is a table showing the data and the resulting four-point moving averages for the car production figures.

Year	Quarter	Cars (1000s)	Four-point moving average
2001	1	52	
	2	50	
			48.25
	3	43	
			47.75
	4	48	
			46.50
2002	1	50	
			47.00
	2	45	
			48.00
	3	45	
			47.75
	4	52	
			48.25
2003	1	49	
			47.00
	2	47	
			47.25
	3	40	
			49.50
	4	53	
			50.50
2004	1	58	
			52.25
	2	51	
			52.50
	3	47	
			51.75
	4	54	
			52.75
2005	1	55	
			52.25
	2	55	
			51.75
	3	45	
	4	52	

> **These values should be lined up between adjacent quarters, e.g. 48.25 should be between Q2 and Q3.**

When the four-point moving averages are plotted on a graph the trend is much clearer.

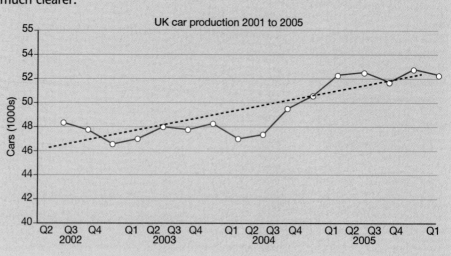

Cumulative frequency

The **cumulative frequency** tells you the number of observations that are less than (<) or less than or equal to (≤) a particular value. The cumulative frequency is calculated by adding each frequency from a frequency table to the sum of its predecessors – a running total of the previous frequencies. The last value will always be equal to the total number of observations, since all frequencies will already have been added to the previous total.

For example, a survey about the number of TV sets in households resulted in this cumulative frequency table.

Number of TV sets	Number of households (frequency)	Cumulative frequency		
0	2	2	←	= 2
1	23	25	←	= 2 + 23
2	67	92	←	= 25 + 67
3	51	143	←	= 92 + 51
4	13	156	←	= 143 + 13
5	3	159	←	= 156 + 3

Using the cumulative frequency column it is easy to see that 92 of the households had two or fewer TV sets. The median is the value half-way through the data. The cumulative total is 159 – there are 159 pieces of data – so the median is the 80th value. At the end of the '1 TV sets' you have reached the 25th value and at the end of the '2 TV sets' you have reached the 92nd value, so the 80th must be '2 TV sets', i.e. the median is 2.

For large sets of data it usually helps to group the raw data.

Example
Here are the ages of people using a skateboard park, recorded as the result of a survey on a Sunday afternoon.

16	5	14	19	14	23	11	13	17	18
14	26	9	15	21	14	20	18	14	20
9	18	21	14	14	16	19	12	12	10
15	12	7	17	14	10	20	13	18	18
10	14	13	14	13	18	13	12	19	10
24	11	10	22	20	11	14	17	11	11
7	19	12							

The results are best put into a grouped frequency table.

A band of 10 years would be too coarse; a band of 2 years too fine. A band of 5 years gives a table of about 6 rows, which is a sensible number.

Age in years, a	Frequency	Cumulative frequency
$0 \le a < 5$	0	0
$5 \le a < 10$	5	5
$10 \le a < 15$	31	36
$15 \le a < 20$	17	53
$20 \le a < 25$	9	62
$25 \le a < 30$	1	63

Reading from the cumulative frequency table, 36 of the skateboard users are less than 15 years old.

The values in the cumulative frequency table can be plotted and the points joined up to give a cumulative frequency curve as shown below.

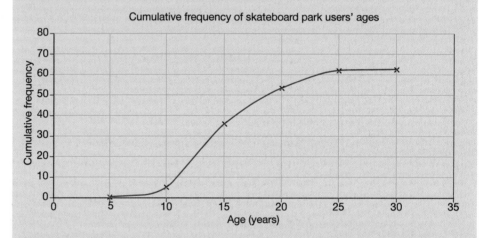

Cumulative frequency of skateboard park users' ages

The median age can be read off from the cumulative frequency curve. There were 63 subjects, so the median age will be the 32nd age. Find the 32nd age on the vertical scale, go across until you reach the curve and read the age that this point corresponds to – the median age. This is about 14 years. Check that you agree.

Quartiles

Quartiles, as the name suggests, are values that divide a sample of data into four groups containing (as far as possible) equal numbers of observations. A data set has three quartiles. References to quartiles sometimes mean the outer two – the **upper and the lower quartiles**; the second quartile is the median.

The lower quartile is the data value a quarter of the way **up** through the ordered data set; the upper quartile is the data value a quarter of the way **down** through the ordered data set (or three-quarters of the way up).

For example: 6   47   49   15   43   41   7   39   43   41   36
when ordered: 6   7   **15**   36   39   **41**   41   43   **43**   47   49
so that: lower quartile = **15**, median = **41**, and upper quartile = **43**.

Inter-Quartile Range (IQR) is a measure of the spread within a data set. It is calculated by taking the difference between the upper and the lower quartiles.

For the case above IQR = (43 – 15) = 28
The IQR is the width of an interval that contains the middle 50% of the sample.

Check that the upper, lower quartiles and IQR for the age data shown in the cumulative frequency curve are about 12 years, 19 years and (19 – 12 =) 7 years respectively.

Box and whisker plots

Box and whisker plots, which are sometimes just called box plots, show how the data is distributed. They show the median, the quartiles, and the smallest and greatest values in the distribution. They can be very useful in giving a quick 'picture' of a distribution and for comparing two distributions.

Example

The marks gained in a
test, in order, are: 18   27   34   52   54   59   61   68   78   82   85   87   91   93   100
The median mark is 68:   18   27   34   52   54   59   61 |**68**| 78   82   85   87   91   93   100
The lower quartile is 52:   18   27   34 |**52**| 54   59   61   68   78   82   85   87   91   93   100
The upper quartile is 87: 18   27   34   52   54   59   61   68   78   82   85 |**87**| 91   93   100
The lowest mark is 18: |**18**| 27   34   52   54   59   61   68   78   82   85   87   91   93   100
The highest mark is 100: 18   27   34   52   54   59   61   68   78   82   85   87   91   93|**100**|

(Had there been an even number of figures in a particular group the quartile would be the mean of the two middle results – exactly the same as in the case of the median for an even number of results.)

This is a box and whisker plot showing the above results.

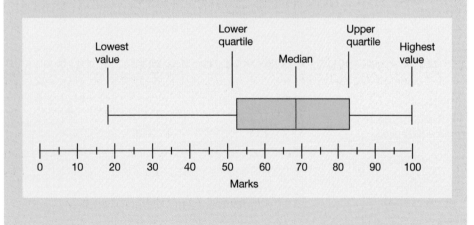

A box plot gives you a good idea as to the shape of the distribution.

- **Symmetrical distribution**. Median in centre, whiskers of equal lengths.

- **Negative skew**. Median closer to upper quartile, whiskers of unequal length.

- **Positive skew**. Median closer to lower quartile, whiskers of unequal length.

Histograms

- A histogram is similar to a bar chart, but it is used to represent **continuous data**.
- With bar charts, the vertical scale shows the actual frequency, but in histograms the vertical scale is labelled **frequency density** so the **area of a bar** is proportional to the frequency of a particular class interval (with bar charts it is the height).
- There are no gaps between the bars of a histogram.
- The bars are drawn between the class boundaries in histograms.

Example

This frequency table shows the lengths of phone calls made over a month.

Length of call in minutes, t	Frequency	Frequency density
$0 \leq t < 5$	20	$20 \div 5 = 4$
$5 \leq t < 10$	35	$35 \div 5 = 7$
$10 \leq t < 15$	40	$40 \div 5 = 8$
$15 \leq t < 20$	25	$25 \div 5 = 5$
$20 \leq t < 25$	15	$15 \div 5 = 3$
$25 \leq t < 30$	10	$10 \div 5 = 2$
$30 \leq t < 35$	5	$5 \div 5 = 1$
$35 \leq t < 40$	3	$3 \div 5 = 0.6$
$40 \leq t < 45$	1	$1 \div 5 = 0.2$
$45 \leq t < 50$	1	$1 \div 5 = 0.2$

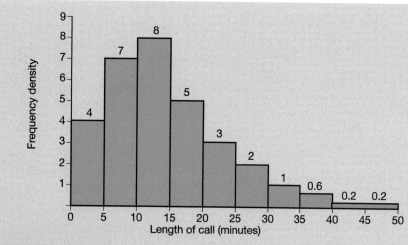

Check for yourself that the areas of the bars give the frequencies given in the table.

Ten bars might be considered too many. Combining bars, giving different class widths, will mean that you need to calculate the frequencies as shown here.

> Frequency density = frequency ÷ class width.

Class times, t, in minutes	Frequency width	Frequency	Frequency density
$0 \leq t < 5$	5	20	$20 \div 5 = 4$
$5 \leq t < 10$	5	35	$35 \div 5 = 7$
$10 \leq t < 15$	5	40	$40 \div 5 = 8$
$15 \leq t < 20$	5	25	$25 \div 5 = 5$
$20 \leq t < 25$	5	15	$15 \div 5 = 3$
$25 \leq t < 35$	10	15	$15 \div 10 = 1.5$
$35 \leq t < 50$	15	5	$5 \div 15 = 0.3$ (1 d.p.)

> Note the different frequency widths in the last two classes.

The above table produces this histogram.

> The unit of frequency density is 'calls per minute'

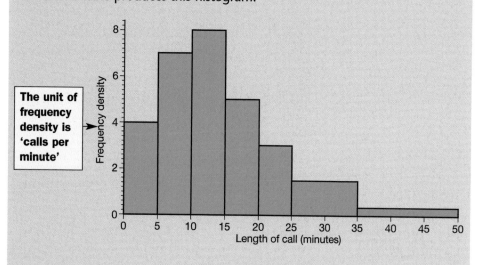

KEY POINT

Histograms often have no label on the vertical axis. It is assumed that the label is frequency density.

It is possible to estimate the mean of a sample from a fully labelled histogram.

Example

The results of an investigation into the lifetimes of personal digital radio batteries are shown in the histogram.

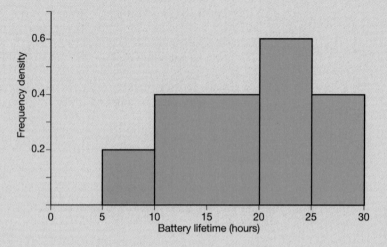

Find the number of batteries tested and estimate the mean battery lifetime.

$$\text{Frequency density} = \frac{\text{frequency}}{\text{column width}}$$

so

frequency = frequency density × column width

The frequencies are therefore:

$$0.2 \times 5 = 1$$
$$0.4 \times 10 = 4$$
$$0.6 \times 5 = 3$$
$$0.4 \times 5 = 2$$

giving a total frequency of 11, so 11 batteries were tested.

The mean can be estimated using the mid-point of each class width.

$$\Sigma fx = (1 \times 7.5) + (4 \times 15) + (3 \times 22.5) + (2 \times 27.5)$$
$$= 190 \text{ (see page 165)}.$$

So $\bar{x}$ (the mean) $= \dfrac{190}{10} = 19$ hours.

1 In order to investigate the improvement, if any, in students' recall of their 'tables', a teacher tested a group before and after an intensive day's practice. Here are the group's 'before and after' scores.

| Score before practice | 2 | 2 | 4 | 8 | 4 | 7 | 9 | 5 | 7 | 6 |
| Score after practice | 5 | 6 | 6 | 9 | 8 | 7 | 10 | 7 | 10 | 8 |

(a) Compare the 'before and after' scores using mean, median and range.
(b) Plot the 'before and after' scores on a scatter graph. What does the graph tell you?

2 This table shows the 1992 UK population (in millions) for different age ranges of young people.

Age in years, a	Population (millions) [frequency]
$0 \leqslant a < 1$	0.7
$1 \leqslant a < 5$	2.7
$5 \leqslant a < 15$	7.5
$15 \leqslant a < 25$	7.9

Draw a histogram showing this information.

3 This table shows the lengths of 250 telephone calls made by a school.

Length of call, t, minutes	Number of calls	Cumulative frequency
$0 < t \leq 4$	24	
$4 < t \leq 8$	47	
$8 < t \leq 12$	68	
$12 < t \leq 16$	44	
$16 < t \leq 20$	32	
$20 < t \leq 24$	21	
$24 < t \leq 28$	8	
$28 < t \leq 32$	6	

(a) Copy the table and complete the cumulative frequency column.
(b) Draw the cumulative frequency curve.
(c) From your graph find:
   (i) the median length of the calls
   (ii) the inter-quartile length of the calls.

4 Use the table in **3** to calculate an estimate of the mean length of a telephone call.

5 This histogram shows the ages of people who live in a small village.

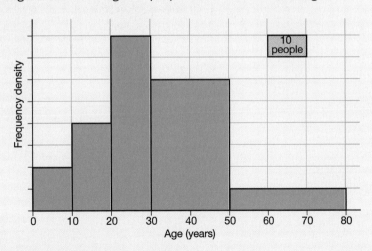

(a) How many people live in the village?
(b) Estimate their mean age.

6 This table shows for each season (winter, spring, summer and autumn) from 1970 to 1972, the number of reported cases of mumps in New York.

	Winter	Spring	Summer	Autumn
1970	412	1371	1369	593
1971	944	1344	876	271
1972	417	771	639	244

(a) Calculate a 4-point moving average.
(b) Draw a graph showing the original data and the moving average.
(c) Draw in the trend line, comment on the trend.

7 This box plot shows the distribution of average monthly temperatures, to the nearest 1 °C, in Oxford for 1855.

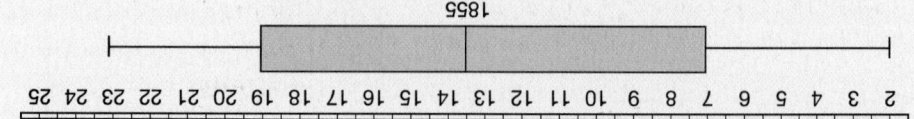

Temperature (degrees Celsius)

The corresponding figures for 2005 are given in this table.

Month	J	F	M	A	M	J	J	A	S	O	N	D
Temperature (°C)	10	7	11	14	17	22	22	23	21	17	11	9

(a) Find the median.
(b) Find the upper and lower quartiles.
(c) Use these values to draw a box plot of monthly temperatures for 2005.
(d) Compare the distribution of monthly temperatures for 1855 and 2005.

2

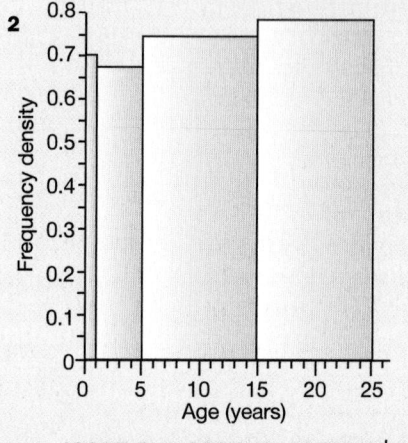

The improvement is pretty even for all students – all improve by about the same.

(b)

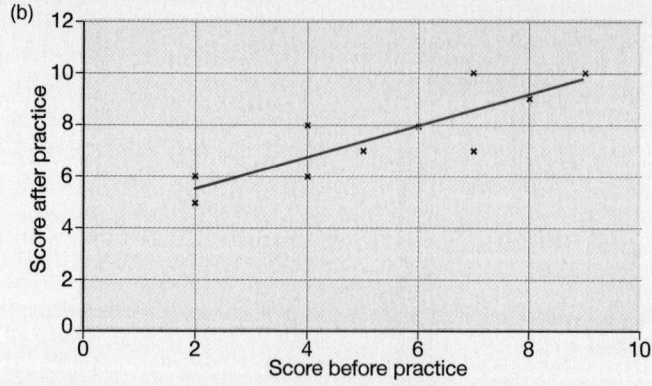

After practice there is improvement in scores.

1 (a)

	Mean	Median	Range
Before practice	5.4	5.5	7
After practice	7.6	7.5	5

3 (a)

Length of call, t, minutes	Number of calls	Cumulative frequency
$0 < t \leqslant 4$	24	24
$4 < t \leqslant 8$	47	71
$8 < t \leqslant 12$	68	139
$12 < t \leqslant 16$	44	183
$16 < t \leqslant 20$	32	215
$20 < t \leqslant 24$	21	236
$24 < t \leqslant 28$	8	244
$28 < t \leqslant 32$	6	250

(b)

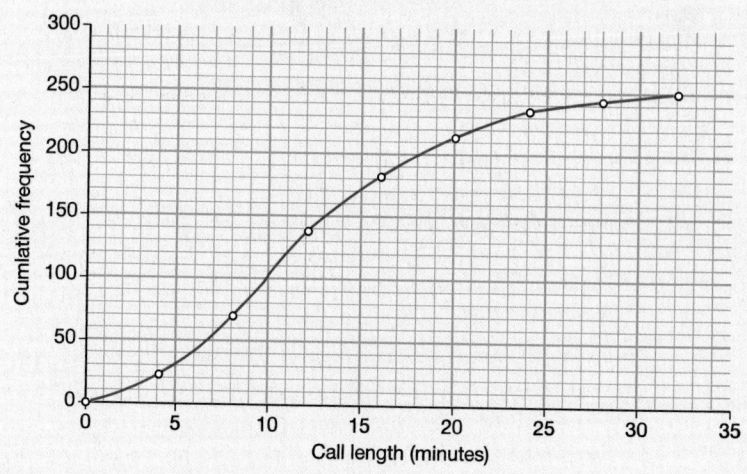

(c) (i) about 11 minutes   (ii) about 8 minutes

4 12.208

5 (a) 290 people   (b) 33 years

6

Original	Four-point average
412	
1371	
1369	936.25
593	1069.25
944	1062.5
1344	939.25
876	858.75
271	727
417	583.75
771	524.5
639	517.75
244	

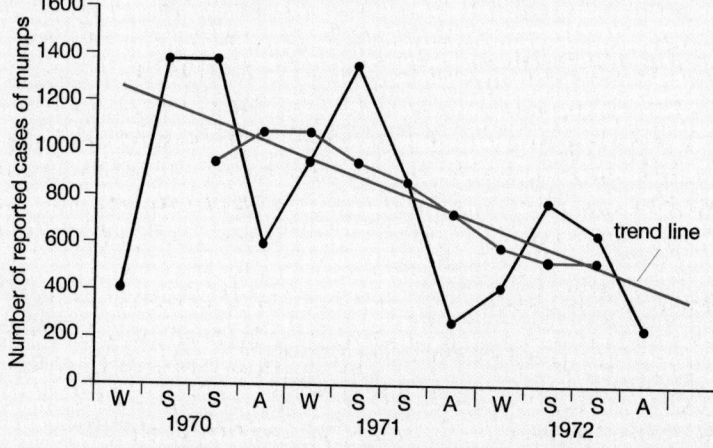

The trend is for the number of reported cases of mumps to decease over time.

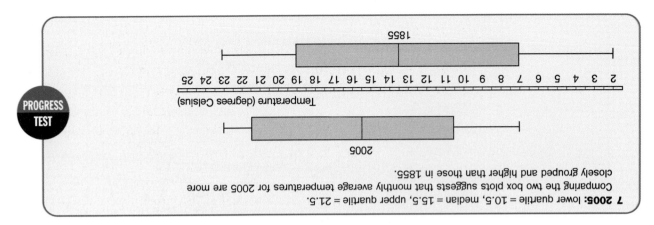

7 2005: lower quartile = 10.5, median = 15.5, upper quartile = 21.5.
Comparing the two box plots suggests that monthly average temperatures for 2005 are more closely grouped and higher than those in 1855.

4.4 Interpreting and examining data

LEARNING SUMMARY

After studying this section, you will be able to:
● *Analyse and compare data presented in a variety of forms*

Comparing data and drawing conclusions

You will need to be able to compare sets of data and draw conclusions about them.

Example

This table shows the average daily maximum temperature in °C for each month in Manchester and Vladivostok.

Month	J	F	M	A	M	J	J	A	S	O	N	D
Vladivostok	–9	–6	1	12	12	18	22	25	21	12	3	–6
Manchester	8	8	10	12	17	19	21	21	18	14	10	8

(a) The mean of the temperatures for Vladivostok is 8.4 °C. Calculate the mean of the temperatures for Manchester.
(b) Work out the range of the temperatures for each city.
(c) Which of the cities has the greater variation in average daily maximum temperatures.

(a) 13.8 °C
(b) Vladivostok: 34 °C   Manchester: 13 °C
(c) Vladivostok has the greater range, so has the greater variation.

Example

A market gardener grows plum trees. He experiments with different fertilisers to see which one gives the best results. The results giving the weights in grams (yield), of the plums from a sample of trees are shown in this table.

Fertiliser A	52	53	45	57	48	49	53	56	58	59	45	62	53
Fertiliser B	57	45	53	52	49	53	48	58	59	56			

Calculate the mean, median, mode and range for each set of data and compare the effectiveness of each fertiliser.

Fertiliser A   **mean = 53.1**   **median = 53**   **mode = 53**   **range = 17**
Fertiliser B   **mean = 53**   **median = 53**   **mode = 53**   **range = 14**

A suitable comment is: Although the mean yield of both fertilisers is similar, the yield for A is slightly better, but the yields using B are more consistent.

Example

Here is a list of test results for two groups of students.

Group A

41	43	46	52	63	69	70	72	76	77	57
74	78	69	55	67	61	58	63	65	74	55

Group B

56	51	72	67	54	67	64	62	56	45	52
58	52	75	74	43	42	50	43	61	45	62

Compare the distributions of the marks for group A and group B.

	Group A	Group B
Mean (1 d.p.)	63.0	56.9
Median	64	56
Mode	63	56
Range	37	33

On three indicators group A's results are better, although given the spread of marks, mode is not a very useful indicator to use here. The results for group A, however, are more spread out than B's (A has the greater range).

A better overall 'feel' for the two sets of marks might be gained by making a 'back to back' stem and leaf table.

```
      Group A    Group B
         631│4│23355
       87552│5│01224668
     9975331│6│122477
     8764420│7│245
```

This shows that scores for group A are higher and tend more towards the higher scores than those for group B.

PROGRESS TEST

1 Amy and Ben are members of a quiz team.
Here are their scores for the previous season.

Amy:	4	8	7	7	5	7	8	7	7	10
	7	7	8	6	7	6	7	7	8	9

Ben:	8	7	7	9	9	6	8	7	9	7
	8	9	8	7	4	7	8	8	9	7
	10	7	8	9	9					

Compare how well each person did last season using, mean, median, mode and range

1.

	Amy	Ben
Mean:	7.1	7.8
Median:	7	8
Mode:	7	7
Range:	6	6

Overall Ben has done better – higher mean and median.

4.5 Probability

LEARNING SUMMARY

After studying this section, you will be able to:
- **understand what is meant by independent and mutually exclusive events**
- **use the multiplication rule for independent events**
- **use the addition rule for mutually exclusive events**
- **draw and use tree diagrams**

Independent events

Two events are independent if the outcome of one of the events gives you no information about whether or not the other event will occur. The events have no influence on each other. For example, the probability of picking a number at random from a list can have no effect on the probability of your getting a double six when you throw two dice.

If two events, A and B, are independent the probability that they both occur is equal to the product of the probabilities of the two individual events, i.e. $P(A \text{ and } B) = P(A) \times P(B)$.

Example
Sanjay has two fair 8-sided dice, both numbered 1 to 8.
He rolls them both. What is the probability that:

(a) both dice show 1
(b) the first dice shows 3 and the second an even number
(c) both dice show multiples of 4
(d) both dice show prime numbers?

What happens on a throw is independent of any previous throws.
So we can use the multiplication rule.

(a) $\dfrac{1}{8} \times \dfrac{1}{8} = \dfrac{1}{64}$

(b) $\dfrac{1}{8} \times \dfrac{1}{2} = \dfrac{1}{16}$

(c) $\dfrac{2}{8} \times \dfrac{2}{8} = \dfrac{4}{64} = \dfrac{1}{16}$

(d) $\dfrac{4}{8} \times \dfrac{4}{8} = \dfrac{16}{64} = \dfrac{1}{4}$

Mutually exclusive events

Events are mutually exclusive if they cannot happen at the same time. For example, if you roll a dice it is impossible to get a 3 at the same time as getting an even number.

For two mutually exclusive events, A and B, the probability that either event A *or* event B will occur is found by adding their probabilities together, i.e. P(A *or* B) = P(A) + P(B).

Example
These cards are turned over and shuffled, then a card is picked at random.

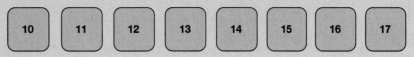

| 10 | 11 | 12 | 13 | 14 | 15 | 16 | 17 |

Event A is 'The number picked is less than 13'.
Event B is 'The number picked is a multiple of 3'.
Event C is 'The number picked is prime.'
Event D is 'The number picked is 13.'
(a) Which of these events are mutually exclusive?
    (i) A, B   (ii) A, D   (iii) B, C   (iv) C, D
(b) (i) What is the probability of picking a number that is either prime or a multiple of 4?
    (ii) What is the probability of picking 10 or 17?

(a) (i) no   (ii) mutually exclusive   (iii) mutually exclusive   (iv) no

(b) (i) $\dfrac{3}{8} + \dfrac{2}{8} = \dfrac{5}{8}$ (ii) $\dfrac{1}{8} + \dfrac{1}{8} = \dfrac{2}{8} = \dfrac{1}{4}$

Tree diagrams

Probability trees can be used to show the outcomes of two or more events. Each branch represents a possible outcome for an event. The probability of each outcome is written on the branch, and the final result depends on the path taken through the tree.

Example

A multiple choice test has five possible answers for each question. Jean finds there are two questions that she has no idea about. So she just guesses and picks each answer at random.

(a) Show all the possible outcomes for these two questions on a tree diagram.

(b) Use the tree diagram to find the probability that Jean guesses:
   (i) both answers correctly   (ii) both answers incorrectly
   (iii) either one of the answers correctly (iv) at least one answer correctly.

The probability of getting the right answer in either case is $\frac{1}{5}$ or 0.2, so the probability of getting the wrong answer is $\frac{4}{5}$ or 0.8.

(Remember the sum of probabilities = 1, so P(wrong) = 1 – P(right).)

(a)

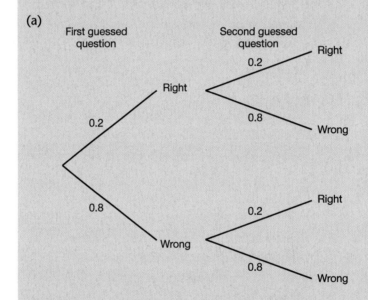

(b) (i) $0.2 \times 0.2 = 0.04$ (ii) $0.8 \times 0.8 = 0.64$

(iii) $0.2 \times 0.8 + 0.8 \times 0.2 = 0.16 + 0.16 = 0.32$

(iv) There are two methods that could be used:

P(at least one right) = 1 – P(both wrong) = 1 – 0.64 = 0.36

or P(at least one right) = P(both correct) + P(right, wrong) + P(wrong, right)

$= 0.2 \times 0.2 + 0.2 \times 0.8 + 0.8 \times 0.2$

$= 0.36$

Questions of the type involving taking marbles are quite common. Read these carefully to see whether or not there is replacement or not of the marbles (or similar objects).

Example

A bag contains 10 marbles of which 2 are red and 8 are green.

A marble is picked at random, its colour is noted and then it is replaced.

This is done three times.

(a) Draw a tree diagram to illustrate all the possible outcomes.

(b) Calculate the probability that exactly two of the three marbles are red.

Now three marbles are picked at random, one after the other, and not replaced.

(c) Calculate the probability that all three marbles are green.

(a)

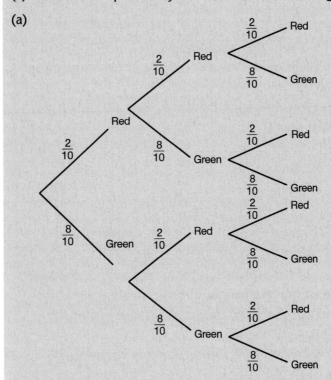

(b) P(red, red, green) + P(red, green, red) + P(green, red, red)

$= (\frac{2}{10} \times \frac{2}{10} \times \frac{8}{10}) + (\frac{2}{10} \times \frac{8}{10} \times \frac{2}{10}) + (\frac{2}{10} \times \frac{8}{10} \times \frac{2}{10}) = \frac{96}{1000}$

(c) Marbles are not replaced: $\frac{8}{10} \times \frac{7}{9} \times \frac{6}{8} = \frac{336}{720} (= \frac{7}{15})$

1 Aijaz has to drive through two sets of traffic lights on his way to work.
These lights work independently.
At the first set the probability he has to stop is 0.6.
At the second set this probability is 0.7.
Calculate the probability that Aijaz is stopped by at least one of these sets of lights. (You may find a tree diagram useful.)

2 An ordinary fair 6-sided dice is thrown three times. What is the probability that there is at least one six?

3 An envelope contains counters of three different colours.
Four are red, ten are blue and six green.
Two counters are to be removed without replacement.
Calculate the probability that both counters will be the same colour.

1 P(stopped at least once) = 1 − P(not stopped at all) = 1 − 0.12 = 0.88 or
P(stopped at least once) = P(stopped, stopped) + P(not stopped, stopped) + P(stopped, not stopped)
= (0.6 × 0.7) + (0.4 × 0.7) × (0.6 × 0.3) = 0.88

2 P(at least one six) = 1 − P(no sixes) = 1 − $(\frac{5}{6})^3$ = $\frac{91}{216}$

3 P(two the same) = P(red, red) + P(blue, blue) + P(green, green)
(Remember there is no replacement in this case.) = $(\frac{4}{20} \times \frac{3}{19}) + (\frac{10}{20} \times \frac{9}{19}) + (\frac{6}{20} \times \frac{5}{19}) = \frac{132}{380} = \frac{33}{95}$

Sample GCSE questions

1 Choy found the play time of each track on her MP3 player.
This table shows the grouped distribution of these times.

Time, t (minutes)	$0 < t \leq 5$	$5 < t \leq 10$	$10 < t \leq 15$	$15 < t \leq 20$	$20 < t \leq 25$	$25 < t \leq 30$
Number of tracks (frequency)	24	86	42	34	10	4

(a) (i) Write down the modal class. **[1]**
(ii) Calculate an estimate of the mean play time of an MP3 track. **[3]**
(b) (i) Complete this cumulative frequency table for the above data. **[2]**

Time, t (minutes)	$t \leq 5$	$t \leq 10$	$t \leq 15$	$t \leq 20$	$t \leq 25$	$t \leq 30$
Number of tracks (frequency)						200

(ii) What is the probability that a randomly chosen track has a play time longer than ten minutes? **[1]**
(c) Draw a cumulative frequency curve showing the play times of the MP3 tracks.
Use your cumulative frequency curve to estimate the median play time of an MP3 track. **[4]**

(a) (i) The modal class is $5 < t \leq 10$ ✔
(ii) Using the mid-points of each interval
$(2.5 \times 24) + (7.5 \times 86) + (12.5 \times 42) + (17.5 \times 34) +$
$(22.5 \times 10) + (27.5 \times 4)$
$= 2160$ ✔
Mean $= 2160 \div 200$ ✔
$= 10.8$ minutes ✔

> Always show your working clearly – you may pick up some method marks.

(b) (i) ✔✔

Time, t (minutes)	$t \leq 5$	$t \leq 10$	$t \leq 15$	$t \leq 20$	$t \leq 25$	$t \leq 30$
Number of tracks (cumulative frequency)	24	110	152	186	196	200

(ii) $(200 - 110) = 90$ tracks are longer than 10 min so the
probability is $\frac{90}{200} = \frac{9}{20}$. ✔

Sample GCSE questions

(c) ✔✔✔

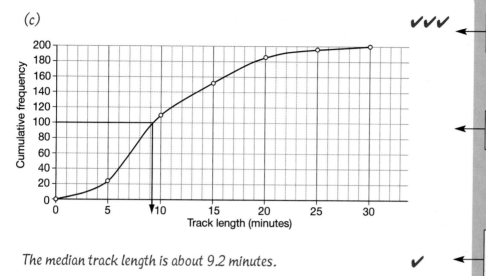

> Join up the points to make a smooth curve.

> Don't forget to label both axes.

The median track length is about 9.2 minutes. ✔

> Show how you used the chart to find the median – you might get some credit for this.

2 Chris flips a drawing pin 100 times. It lands 'pin up' 85 times.

(a) What is the probability of the drawing pin landing 'pin down'? **[2]**

(b) Chris drops the pin three times in succession. Calculate the probability that:

(i) all three land 'pin down' **[2]**

(ii) at least one lands 'pin up'. **[3]**

(a) $P(pin\ down) = 1 - P(pin\ up) = 1 - \frac{85}{100}$ ✔

$= \frac{15}{100} = 0.15$ ✔

> P(event) = 1 – P(not event) is very useful.

(b) (i) $P(pin\ down, pin\ down, pin\ down)$

or $P(d, d, d) = P(pin\ down) \times P(pin\ down) \times P(pin\ down)$

$= 0.15 \times 0.15 \times 0.15$ ✔

$= 0.003\,375$ ✔

> Make your method clear – you may get partial credit even if there is a slip in your calculation.

(ii) $P(at\ least\ 1\ pin\ up) = P(u, u, u) + P(u, u, d) + P(u, d, u)$

$+ P(d, u, u) + P(u, d, d) + P(d, u, d) + P(d, d, u)$ ✔

$= 0.85^3 + 3 \times 0.85 \times 0.85 \times 0.15 +$

$3 \times 0.85 \times 0.15 \times 0.15$ ✔

$= 0.996\,625$ ✔

or $P(at\ least\ 1\ pin\ up) = 1 - P(d, d, d)$

$= 1 - 0.003\,375 = 0.996\,625$

3 If it is fine today the probability that it will rain tomorrow is $\frac{1}{5}$. If it is rainy today the probability that it will rain tomorrow is $\frac{2}{3}$.
The probability that today is fine is $\frac{1}{2}$.

(a) Complete this tree diagram using the information given above.

Sample GCSE questions

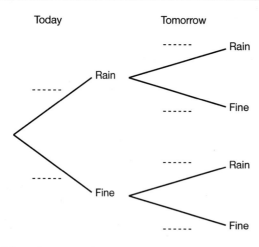

Today   Tomorrow

    Rain

    Rain

    Fine

    Rain

    Fine

    Fine   **[3]**

(b)   Use the tree diagram to find the probabilities that:

(i)   it will be fine on both days   **[1]**

(ii)   that there will be different weather today and tomorrow.   **[2]**

(a)

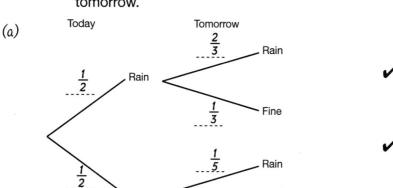

Today   Tomorrow

$\frac{1}{2}$ Rain $\frac{2}{3}$ Rain ✔

    $\frac{1}{3}$ Fine

$\frac{1}{2}$ Fine $\frac{1}{5}$ Rain ✔

    $\frac{4}{5}$ Fine ✔

✔

Always check that the sum of the probabilities from a branch add to 1.

(b)   (i) $P(fine, fine) = \frac{1}{2} \times \frac{4}{5} = \frac{2}{5}$ ✔

     (ii) $P(rain, fine) + P(fine, rain)$

$$= \left(\frac{1}{2} \times \frac{1}{3}\right) + \left(\frac{1}{2} \times \frac{1}{5}\right)$$ ✔

$$= \frac{1}{6} + \frac{1}{10} = \frac{4}{15}$$ ✔

Always be on the look out for more than one route through the probability tree.

4   This table shows the distribution of adults' ages in the UK for the over 25s.

Age, x years	Number of people (millions)
$24 < x \le 34$	8.0
$34 < x \le 59$	20.5
$59 < x \le 74$	7.8
$74 < x \le 100$	4.5

Sample GCSE questions

(a) Draw a histogram to show this distribution. **[3]**
(Work to 2 s.f. accuracy.)

(b) A supermarket wants to choose a stratified random sample of 1000 over 25s in the UK. Show how this can be done. **[3]**

(a)

Age, x years	Number of people (millions)	Frequency density (to 2 s.f.)
$24 < x \leqslant 34$	8.0	0.80
$34 < x \leqslant 59$	20.5	0.82
$59 < x \leqslant 74$	7.8	0.52
$74 < x \leqslant 100$	4.5	0.17

> Remember frequency density =
> $$\frac{(\text{frequency of items in the bar's class interval})}{(\text{width of the class interval})}$$

✔

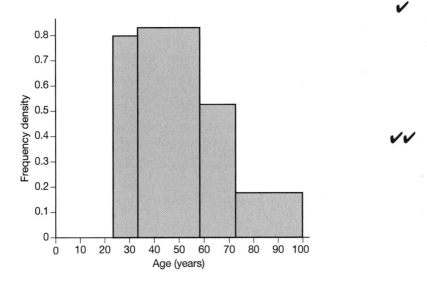

✔✔

(b) The total number of subjects $= 8.0 + 20.5 + 7.8 + 4.5 = 40.8$ ✔
so the numbers required in each age group are:

Age, x years	Number of people (millions)	Number to sample to nearest whole number
$24 < x \leqslant 34$	8.0	$\frac{8.0}{40.8} \times 1000 \approx 196$
$34 < x \leqslant 59$	20.5	$\frac{20.5}{40.8} \times 1000 \approx 502$
$59 < x \leqslant 74$	7.8	$\frac{7.8}{40.8} \times 1000 \approx 191$
$74 < x \leqslant 100$	4.5	$\frac{4.5}{40.8} \times 1000 \approx 110$

> Lay out your working logically – it will prevent your making simple slips.

 ✔✔

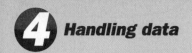

Exam practice questions

1 This table shows the distribution of the working life of 22 PC projector lamps.

Working life, L, thousands of hours	Number of lamps
$1.6 \leq L < 1.8$	2
$1.8 \leq L < 2.0$	4
$2.0 \leq L < 2.1$	8
$2.1 \leq L < 2.2$	4
$2.2 \leq L < 2.6$	4

(a) (i) Construct a histogram to display this data. **[3]**

    (ii) Calculate an estimate of the mean working life of the PC projector lamps. **[4]**

(b) A new supplier of PC projector lamps produces this data on some tests on
100 Mark A and 100 Mark B projector lamps. The units are thousands of hours.

	Minimum	Lower quartile	Median	Upper quartile	Maximum
Mark A	1.5	2.5	3.2	3.8	4.6
Mark B	1.3	2.3	3.4	4.1	4.9

(i) Draw two box plots for the working lives of the Mark A and Mark B lamps. **[3]**

(ii) Compare the working lives of the Mark A and Mark B lamps. **[2]**

2 The probability of a train arriving early at a station is 0.05.
The probably of a train arriving late at the station is 0.1.

(a) During a week 400 trains are expected at the station.
How many are likely to arrive at the correct time? **[3]**

(b) (i) What is the probability that two trains in succession are late? **[2]**

    (ii) What crucial assumption have you made in your answer to (ii)? **[1]**

3 Some students were asked to estimate the height of a building in metres.
Here are the results:

```
8    9    10   10   10   10   10   10   11   11   11   11   12
12   13   13   13   14   14   14   15   15   15   15   15   15
15   15   16   16   16   17   17   17   17   18   18   20   22
25   27   35   38   40
```

(a) Plot the results on a stem and leaf diagram. **[3]**

(b) The true height was 13 m. What can you say about the students' estimation skills? **[1]**

Exam practice questions

4 The table below shows the chirping rate per second of a cricket and the temperature in °C.

Chirps per second	20	16	20	18	17	16	15	17	15	16	15	17	16	17	14
Temperature (°C)	31	22	34	29	27	24	21	28	21	29	26	28	27	29	25

(a) Plot the results on a scatter graph. [3]

(b) What does your scatter graph tell you about the correlation between the temperature and the chirping rate of crickets? [1]

5 In his socks drawer a man has 8 black socks and 11 white socks.
He picks two at random. Find the probabilities that:

(a) both are white socks [2]

(b) both are white or both are black [2]

(c) one is white and one is black. [2]

6 This table shows taxable income and number of people who are in each taxable income band in 2005/6 in the UK.

Taxable income	Number of taxpayers (millions)
£4895–£4999	0.1
£5000–£7499	2.9
£7500–£9999	3.5
£10 000–£14 999	6.1
£15 000–£19 999	5.1
£20 000–£29 999	6.4
£30 000–£49 999	4.3
£50 000–£99 999	1.5
£100 000 and over	0.5
All incomes	**30.4**

(a) Write down the modal taxable income class. [1]

(b) Estimate the median taxable income. [3]

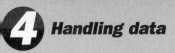
Exam practice questions

7 This graph shows the number of visitors to a small seaside town each quarter.
(The first quarter is Jan, Feb, March and so on.)

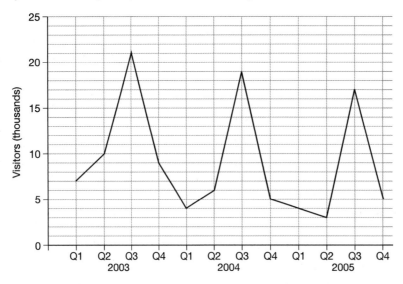

Year	Quarter	Visitors	Four point moving average
2003	1	7	
	2	10	←
	3	21	←
	4	9	←
2004	1	4	←
	2	6	←
	3	19	←
	4	5	←
2005	1	4	←
	2	3	←
	3	17	←
	4	5	

(a) Copy the table and calculate the 4-point moving averages for this data and enter the values on the table. **[3]**

(b) Plot these moving averages on the graph. **[2]**

(c) Draw the trend line. What is the trend over time for the number of visitors to the town? **[2]**

Exam practice answers

Unit 1

1 (a) $72 = 2 \times 2 \times 2 \times 3 \times 3$ or $2^3 \times 3^2$ **[3]**

(b) $72 = 2^3 \times 3^2$ and $48 = 2^4 \times 3$
so LCM $= 2^4 \times 3^2 = 144$ **[2]**

(c) (i) $16^{-\frac{1}{2}} \times 27^{\frac{2}{3}} = \frac{1}{4} \times 3 \times 3 = \frac{9}{4}$ **[2]**

(ii) $(\sqrt[3]{8})^2 = (2)^2 = 4$ **[2]**

With questions such as this there is always a temptation not to show any working and so risking losing any method marks available. In part (a) there would be some credit for a clear attempt to use a factor tree or other method to find the prime factors. In (b) splitting the 48 into its prime factors successfully would get 1 mark. In (c) part (i), there would be 1 mark for working out $16^{-\frac{1}{2}}$ or $27^{\frac{2}{3}}$ correctly. In part (ii) there would be 1 mark for $\sqrt[3]{8} = 2$.

2 (a) $= 5\frac{1}{2} + 4\frac{1}{3} + 3\frac{1}{5} = (5 + 4 + 3) + \frac{1}{2} + \frac{1}{3} + \frac{1}{5}$ **[1]**

$= 12 + \frac{15 + 10 + 6}{30}$ **[1]**

$= 13\frac{1}{30}$ **[1]**

(b) $= 7\frac{2}{5} - 6\frac{5}{8} = (7 - 6) + \frac{2}{5} - \frac{5}{8}$ **[1]**

$= 1 + \frac{16 - 25}{40}$ **[1]**

$= 1 - \frac{9}{40} = \frac{31}{40}$ **[1]**

(c) $= 1\frac{2}{3} \div 2\frac{2}{9} = \frac{5}{3} \div \frac{20}{9}$ **[1]**

$= \frac{5}{3} \times \frac{9}{20}$ **[1]**

$= \frac{5}{3} \times \frac{9}{20} = \frac{1\cancel{5}}{\cancel{3}1} \times \frac{\cancel{9}3}{\cancel{20}4} = \frac{3}{4}$ **[1]**

These are good examples of the type of 'pure' fraction questions you might be asked. When dealing with mixed fractions (whole number with fraction) take care with subtractions, you can subtract the whole numbers and fractions separately but you may get a negative fraction – this isn't a problem providing you remember this; see part (b). When dividing or multiplying with mixed fractions changing to 'top heavy' fractions before you do anything is a good rule to follow. Full marks are awarded only if answers are given in their simplest terms.

3 $\frac{10}{2 \times 10^{-23}}$ **[2]**

$= 5 \times 10^{23}$ **[1]**

Candidates often make unnecessary errors in questions of this type because of the presence of indices and unfamiliar numbers. This can be avoided if you think through the problem with simpler numbers or even algebra, for example if carbon atoms weighed 2 g (!) or w grams there would be $10 \div 2$ ($10 \div w$) or $\frac{10}{2}$ or $\frac{10}{w}$ in 10 g of pure carbon. After this the only other possible difficulty is recalling that $\frac{1}{10^{-27}} = 10^{27}$

4 $(2 - \sqrt{3})^2 = (2 - \sqrt{3}) \times (2 - \sqrt{3})$
$= 4 - 2\sqrt{3} - 2\sqrt{3} + (^-\sqrt{3}) \times (^-\sqrt{3})$ **[1]**
$= 4 - 4\sqrt{3} + 3 = 7 - 4\sqrt{3}$
so $a = 7$ and $b = -4$ **[1]**

A pretty straightforward question, but made more challenging by the presence of negative quantities, for example $-\sqrt{3} \times -\sqrt{3}$. It is better to take another line of working rather than trying to juggle several arithmetic expressions in your head at the same time – by using another line you reduce the scope for error and make your thinking clearer to the examiner. Some candidates have difficulty with the signs attached to a and b in $a + b\sqrt{3}$ for $7 - 4\sqrt{3}$, but if you look carefully, $a = 7$ and $b = -4$.

5 After a bounce the height $= 0.8 \times$ height of previous bounce, **[1]**
so after 3rd bounce
height $= 0.8 \times [0.8 \times (0.8 \times 250)]$ **[1]**
$= 0.8^3 \times 250$ **[1]**
$= 128$ cm **[1]**

The key to this question and to other similar questions involving repetitive increases or decreases (including compound interest) is to find the scale factor. In this case there is a 20% decease in height on the previous bounce – so the scale factor after each bounce is 0.8. Clear evidence that you realise this will gain partial credit (if it had increased by 20% in height each bounce then the scale factor would have been 1.2). Although this was a 'non-calculator' question (you should be able to work out $0.8^3 \times 250$ using pencil and paper), this type of question usually appears on the calculator section.

6 (a) Lower bound $= 17.5$ kg **[1]**
and upper bound $= 18.5$ kg **[1]**

(b) Upper and lower bounds for the airport scales are 17.55 and 17.65, respectively. **[1]**
So heaviest (heaviest at start – lightest at end), which is $18.5 - 17.55 = 0.95$ kg **[1]**

Exam practice answers

An upper bound of 18.499 999 999 would not gain full credit – it is 18.5. The other fact that candidates tend to forget is that the largest result in a subtraction is 'largest possible' – 'smallest possible' and vice versa for the smallest result. In the case of a division 'largest result' = 'largest result' ÷ 'smallest result' and so on. You can always check these simply as such questions almost always occur on the calculator paper.

7 $\sqrt[3]{\dfrac{(24 \times 3600)^2 \times 9.8 \times (6.4 \times 10^6)^2}{4\pi^2}} - 6.4 \times 10^3$

$\sqrt[3]{\dfrac{7.466\,496 \times 10^9 \times 9.8 \times 4.096 \times 10^{13}}{4\pi^2}} - 6.4 \times 10^3$ **[2]**

$= 4.233\,363\,935 \ldots \times 10^7\,\text{m}$ **[1]**

$= 4.2 \times 10^4\,\text{km (to 2 s.f.)}$ **[1]**

It is essential to calculate this expression twice in order to check your answer; it also illustrates how important it is to 'know your own calculator'. The other areas where slips may occur, especial after spending several minutes evaluating a complex expression, is forgetting that the answer is asked for in km and to 2 s.f. – something that is easy to do in the heat of the moment. A quick re-read of the question before going on to the next one can prevent a needless loss of marks. It is a good question to check using approximation

$\sqrt[3]{\dfrac{(24 \times 3600)^2 \times 9.8 \times (6.4 \times 10^6)^2}{4\pi^2}}$

$\approx \sqrt[3]{\dfrac{(8 \times 10^4)^2 \times 10 \times (6 \times 10^6)^2}{4 \times 10}}$

$\approx \sqrt[3]{6 \times 10^{22}} = \sqrt[3]{60 \times 10^{21}} = 10^7 \times \sqrt[3]{60}$

$\approx 4 \times 10^7\,\text{m}$ ignoring the 6.4×10^3 which is too small to affect the size of the final answer. This suggests that the above answer is at least reasonable.

8 117.5% of pre-VAT cost = £850 **[1]**

so 100% of pre-VAT cost $= 100 \times \dfrac{850}{117.5}$ **[1]**

$= £723.40$ to the nearest p **[1]**

Each exam session a large number of candidates merely subtract 17.5% of £850 from £850 or similar and score zero. To avoid this, think of the original cost as 100%, so the new cost is 117.5% which is £850 – from this find 1% (by dividing by 117.5) then multiply by 100 to find 100%, which is the answer to the problem. A good check is to increase the answer by 17.5% to see if you get £850.

Unit 2

1 (a) $4a - 7 = -20$

$4a = -13$ **[1]**

$a = \frac{-13}{4} = -3\frac{1}{4}$ **[1]**

(b) $5(2x - 1) + 6x = 7 - 8x$

$10x - 5 + 6x = 7 - 8x$ **[1]**

$24x = 12$ **[1]**

$x = \frac{1}{2}$ **[1]**

(c) $\dfrac{2}{x} = 4$

$2 = 4x$ **[1]**

$x = \frac{1}{2}$ **[1]**

As in all cases – lay out your working clearly, don't be afraid to take an extra line if it makes your working (and thinking) crystal clear. Part (a) would be worth checking – just in case – as negative solutions involving fractions are not generally common (although all the solutions in this question are in fact fractions). Questions like part (c), involving a simple fraction, quite often catch candidates out, with '8' a common wrong answer – a result perhaps of solving the equation by inspection rather than working it out by multiplying both sides by x (to give $2 = 4x$) in order to remove the fraction. As with any equation check by substituting your solution back in to the original equation – does it fit?

2 $3x - 2y = 10$ (1)

$5x + 4y = 2$ (2)

$(1) \times 2 + (2)$

$6x - 4y = 20$ **[1]**

$5x + 4y = 2$

$11x = 22$

$x = 2$ **[1]**

substituting for x in (1) gives $6 - 2y = 10$,

so $y = -2$. **[1]**

The important words here are 'algebraically' and 'show your working': it means don't use trial and improvement – use algebra. Solutions found by trial and improvement or merely by spotting will gain little credit. Always number your equations and make your intentions clear – do not cross out work, you may gain method marks even though your final answer was wrong. If you have sufficient time, check your solutions by substituting back into the original equations. They are usually fairly simple equations so this isn't normally too hard or time consuming – and may gain marks!

3 (a) $L \propto \dfrac{1}{f}$ (L = length, f = frequency)

so $L = k\dfrac{1}{f}$, $k = Lf$, $k = 65 \times 147$ **[1]**

$$L = \frac{65 \times 147}{110}, k = 86.9 \text{ cm (to 3 s.f.)} \quad [1]$$

(b) $f \propto \sqrt{T}$ (f = frequency, T = tension),
so $f = k\sqrt{T}$ [1]
$196 = k\sqrt{T}$ so $k = \dfrac{196}{\sqrt{T}}$

When T doubled, $f = k\sqrt{2T} = \dfrac{196}{\sqrt{T}} \times \sqrt{2T}$

$f = 196 \times \sqrt{2} = 277 \text{ Hz (to 3 s.f.)}$ [1]

Don't calculate until the last line: there might be some cancelling to be done. Remember that '$\propto$' can be always be replaced by ' $= k$', where k is a constant.

4 (a) Circle has radius $\sqrt{25} = 5$ cm, so line has x-intercept 5 and y-intercept 5. Using the diagram, the x-intercept is positive and the y-intercept is positive. [1]
The gradient is 1, y-intercept is 5, so equation must be $y = x + 5$ [1]

(b) $x^2 + y^2 = 25$ (1) and $y = x + 1$ (2)
Substituting $y = x + 1$ from (1) into (2) gives:
$x^2 + (x + 1)^2 = 25$ [1]
Expanding: $x^2 + x^2 + 2x + 1 = 25$
$2x^2 + 2x - 24 = 0$ [1]
$x^2 + x - 12 = 0$
$(x + 4)(x - 3) = 0$
so $x = -4$ or $x = 3$ [1]
Substituting back into (2) for x gives
$y = -3$ or $y = 4$
so the points are (3, 4) and (−3, −4) [2]

(An intercept is where a line (often a straight line) crosses either the x-axis or the y-axis.)
The examiner is really asking for the solutions to the simultaneous equations $x^2 + y^2 = 25$ and $y = x + 1$.
Remember $x^2 + y^2 = n$ is the equation of a circle radius $\sqrt{n}$. Strictly speaking, the x and y intercepts are ± 5. With this type of question the final quadratic might be solved by either factorisation or use of the formula. It will be factorisation if it is on the non-calculator section/ paper, but *may* be the formula if it is on the calculator section/paper.

5 The next integer after n is $n + 1$ [1]
Two consecutive numbers squared are n^2 and $(n + 1)^2$
which are n^2 and $n^2 + 2n + 1$ [1]
The difference is $(n^2 + 2n + 1) - (n^2)$ which is $2n + 1$ [1]
But double any integer always gives an even number ($2 \times$ an integer must be divisible by 2), and 1 added to an even number always gives an odd number, so $2n + 1$ is always odd. [1]

A fairly straightforward example of an algebraic proof. It was made easier by the simple lead-in about two consecutive numbers – you may not always be given this hint. You also need to know that for any integer n, $2n$ is even and $2n + 1$ is odd – a lot of simple algebraic proofs use this.

6
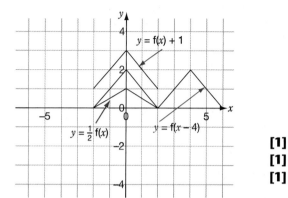
[1]
[1]
[1]

Transformation of graphs is a topic on which many candidates gain few marks, possibly because the obvious answers are wrong – for example $f(x - 4)$ is very often wrongly thought of as being a translation of −4 (i.e. 4 units in the negative direction) – you need to learn these – with the aid of a graph plotter software/ calculator to convince yourself.

7 (a) $x^2 + 6x + 11 = (x + 3)(x + 3) + 2$ [1]
so $a = 3$ and $b = 2$ [2]

(b) (i) $u_n = n^2$ [1]
(ii) $u_n = n^2 - 1$ [1]
(iii) $u_n = n^2 + n$ [1]
(iv) $u_n = 3 \times 2^{n-1}$ [2]

Part (a) is a fairly standard way of asking you to complete the square for the given expression. This is provably best done by inspection – $(x + 3)^2$ gives the x^2 and $6x$ term and then adjust to make the pure number part 11.
In part (b), (i) is the clue to think in terms of squares (you need to be able to recognise square and cubes of integers), parts (ii) and (iii) are perhaps easiest to solve if you write the squares of the first few integers underneath and so what adjustments need to be made, for example:

1	4	9	16	...	n^2
2	6	12	20	...	?

$= 1 + 1, 4 + 2, 9 + 3, 16 + 4, ... n^2 + n,$
$? = n^2 + n$

Part (iv) is probably about as difficult as you are ever likely to be asked. Again, a useful ploy is to line the sequence up and try to see a pattern – like this:

3	6	12	24	48	...
3×1	3×2	3×4	3×8	3×16	...
3×1	3×2	3×2^2	3×2^3	3×2^4	... $3 \times 2^{n-1}$

8 $\dfrac{3}{2x-1} + \dfrac{2}{x+2} = 1$

$\dfrac{3}{2x-1} + \dfrac{2}{x+2} = \dfrac{3(x+2)+2(2x-1)}{(2x-1)(x+2)}$ **[2]**

Which simplifies to $\dfrac{7x+4}{(2x-1)(x+2)}$ **[1]**

So $\dfrac{7x+4}{(2x-1)(x+2)} = 1$ **[1]**

$\therefore 7x + 4 = 2x^2 + 3x - 2$ **[1]**
so $2x^2 - 4x - 6 = 0$
giving $x^2 - 2x - 3 = 0$
$(x-3)(x+1) = 0$ **[1]**
$\therefore \qquad x = 3 \text{ or } x = -1$ **[1]**

As a general rule you need to remove the fractions as soon as possible, in this case by multiplying through by both denominators. If you are in doubt, try the operation with simple numbers, e.g.
$\dfrac{2}{7} + \dfrac{3}{5} = \dfrac{(2 \times 5) + (3 \times 7)}{7 \times 5}$ then the expression is

made more manageable by cross-multiplying to give $7x + 4 = 2x^2 + 3x - 2$. Show each step clearly, so that the examiner will be able to award method marks, of which there are a lot on this question – sounds obvious but marks can only be awarded when there is clear evidence!

9 $\qquad 2n^2 - n = 45$ **[1]**
so $\quad 2n^2 - n - 45 = 0$
$\therefore (2n+9)(n-5) = 0$ **[1]**
$n = -4.5 \text{ or } n = 5$ **[1]**
A negative value makes no sense here, so the answer is the 5th hexagonal number (u_5). **[1]**

A fairly straightforward question, involving setting up and solving a quadratic equation. Nevertheless, candidates tend to lose marks by giving roots in the case of, say, $(x+1)(x-7)$ as 1 and –7 – forgetting that the roots make the value of the bracket zero so they must be –1 and +7 in this case. Sometimes one of the roots can be rejected as a solution because it makes no sense in the question context – like here, having a negative term of a sequence.

Unit 3

1 (a) (i)

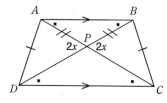

Because $ABCD$ is isosceles, ABP is isosceles.
$\therefore$ angle ABD = angle BAC **[1]**

angle BAC = angle $ACD = x$ (alternate angles) **[1]**
$\therefore$ angle $ABD = x$
(ii) Triangle DCP is also isosceles, so angle $APD = 2x$ (exterior angle of a triangle) **[1]**

(b) angle DAB = angle ABC (symmetry) **[1]**
angle ABC + angle $BCD = 180°$ (AB and DC are parallel) **[1]**
$\therefore$ angle DAB + angle $BCD = 180°$ **[1]**
$\therefore ABCD$ is a cyclic quadrilateral (opposite angles add up to 180°) **[1]**
An isosceles trapezium has one line of symmetry.
It can be very helpful to mark in equal angles and sides. Don't forget, try to give each statement a reason.

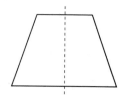

2 (a) $\overrightarrow{XN} = \tfrac{1}{2}\overrightarrow{XY}$ and $\overrightarrow{XY} = \overrightarrow{OB} = \mathbf{b}$
$\therefore \overrightarrow{XN} = \tfrac{1}{2}\mathbf{b}$ **[1]**
$\overrightarrow{ON} = \overrightarrow{OX} + \overrightarrow{XN} = \tfrac{1}{2}\mathbf{a} + \tfrac{1}{2}\mathbf{b}$ **[1]**

$\overrightarrow{AN} = \overrightarrow{AX} + \overrightarrow{XN} = -\tfrac{1}{2}\mathbf{a} + \tfrac{1}{2}\mathbf{b} = \tfrac{1}{2}\mathbf{b} - \tfrac{1}{2}\mathbf{a}$ **[1]**
$\overrightarrow{NB} = \overrightarrow{NY} + \overrightarrow{YB} = \tfrac{1}{2}\mathbf{b} - \tfrac{1}{2}\mathbf{a}$ **[1]**

(b) $\overrightarrow{AB} = \overrightarrow{AC} + \overrightarrow{CB} = \mathbf{b} - \mathbf{a}$ **[1]**
$\overrightarrow{AN} = \tfrac{1}{2}(\mathbf{b} - \mathbf{a})$ and $\overrightarrow{AB} = (\mathbf{b} - \mathbf{a})$
So A, N, B all are on the same straight line, with N as the mid-point **[1]**

This is a popular type of vector question – look out for negative vectors.

3 Any formulae for volume must have units of $(\text{length})^3$. **[1]**
The only one that fits is $2\pi^2 Rr^2$. **[1]**

Dimensions questions are usually pretty straightforward. The main errors occur when candidates assume that π has dimensions of length. Always look out for relationships that are not consistent dimensionally, e.g. two terms, one having the correct dimensions, the other not.

4 Need to find the vertical height of the tetrahedron to find the volume.
From Pythagoras:

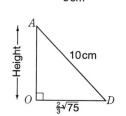

$DM^2 = 10^2 - 5^2 = 75$
$DM = \sqrt{75}$
so $DO = \tfrac{2}{3}\sqrt{75}$

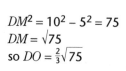

[1]

$(\text{Height})^2 = 10^2 - OD^2$ (Pythagoras) **[1]**
so Height = $\sqrt{\frac{200}{3}}$ **[1]**
Now need to find area of base DCB.
DCB is an equilateral triangle: all angles 60°.
Using the rule: $A(\text{area}) = \frac{1}{2}ab \sin C$
Area of base $= \frac{1}{2}(10 \times 10 \times \sin 60°) = 50\sin 60°$ **[1]**

Using Volume $= \frac{1}{3} \times$ base area $\times$ perpendicular
height gives for the volume
$\frac{1}{3} \times (50\sin 60°) \times (\sqrt{\frac{200}{3}})$ **[1]**
$= 117.9\,\text{cm}^3$ (to 3 s.f.) **[1]**

Always sketch the triangles under consideration: it
helps to clarify your thinking. Check that right
angles really are right angles: 3D drawings can be
deceptive. As a general rule, delay performing the
actual calculation as late as possible; it might just
happen that some numbers cancel in the end.
Follow-through marks would be available in this
multi-step problem. This means that if you made a
mistake in your calculation in an early part of the
solution, as long as you used your wrong answer
correctly later on no further loss of marks would
arise.

5 (a) **[2]**

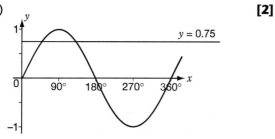

(b) (0, 0), (180°, 0) (360°, 0) **[1]**

(c) $\sin x = 0.75$ **[1]**
Don't forget there are two solutions, as you
can see from the graph above.

(d) Solutions are 48.6° (to 3 s.f.) and 131.4°
(180° − 48.6°) **[2]**

You need to be able to recall the graphs of all three
trigonometric functions and to solve
trigonometrical equations such as these. A sketch is
useful in these cases in any case as it indicates
where the solutions roughly lie and if there is more
than one solution – candidates quite often give
only one solution and fail to gain full credit because
of this momentary slip.

6 (a) Rotation **[1]**
of 180° **[1]**
about (−1, 0) **[1]**

(b) Reflection in the x-axis (or line $y = 0$) **[1]**

(c) Reflection **[1]**
in the line $x = -1$ **[1]**

Transformations must be described in full to gain
full credit, e.g. rotations must have a centre, angle
and direction, reflections a mirror line,
enlargements a centre and a scale factor. In
(c) there must be a single transformation – a
common mistake is for candidates to give a
combination of transformations.

7

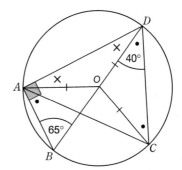

(a) Angle $BAC =$ angle $BDC = 40°$ (angles
subtended by same chord) **[1]**

(b) Angle $BAD = 90°$ (angles subtended by
diameter) **[1]**
angle $DAC = 90° - 40° = 50°$ **[1]**

(c) Angle $AOD = 180° - 25° - 25° = 130°$ **[1]**
(In triangle ABD, angle $ABB = 180° -$
$90° - 65° = 25°$ and AOD is an isosceles
triangle) **[1]**

(d) In triangle AOC, angle $AOC =$ angle $AOB +$
angle BOC
angle $BOC = 2 \times 40°$ (angle at centre is double
angle at the circumference) **[1]**
angle $AOB = 2 \times$ angle $ADO = 2 \times 25°$
(vertically opposite angles) **[1]**
So angle $AOC = 80° + 50° = 130°$
∴ angle $CAO = (180 - 130) \div 2 = 25°$ **[1]**

Make a rough sketch and mark in equal angles and
angles you know– then look back to see what
angles you've been asked to find. Try to use the
rule 'no statement without a reason' as far as you
are able. In some cases an angle, even if correct,
will not gain credit unless accompanied by a
reason. There are several other ways to solve some
parts of the question, for example in (d):
BAD is a right angle, angle $BAC =$ angle $BDC = 40°$,
angle $OAD =$ angle $ODA = 25°$, so angle $OAC = 90°$
$- 40° - 25° = 25°$.

8 Volume of a cylinder height h and radius r is $\pi r^2 h$.
The cylinder of toothpaste has volume
$\pi 0.4^2 \times 2\,\text{cm}^3$ **[1]**
He uses two of these a day, so each day he
uses $0.64\pi\,\text{cm}^3$ **[1]**
So the tube will last $100 \div 0.64\pi$ days **[1]**
$= 49$ days (need to round down) **[1]**

A simple question, but you are given a mixture of units: mm and cm – convert to the same unit as soon as possible. For the context here you would need to round the answer down.

9 Using same units
Volume of steel = volume of ball – volume of spherical hollow inside [1]

Volume of ball = $\frac{4}{3}\pi r^3 = \frac{4}{3}\pi 4^3 = \frac{256 \times \pi}{3}$ cm^3 [1]

Volume of space inside = $\frac{4}{3}\pi r^3 = \frac{4}{3}\pi 3.5^3$

= $\frac{171.5 \times \pi}{3}$ cm^3 [1]

Giving for the volume of steel

$(\frac{256 \times \pi}{3} - \frac{171.5 \times \pi}{3}) = 88.5$ cm^3 (to 3 s.f.) [1]

A fairly straightforward question once you 'see' what is going on. As with any question involving circles or spheres, take care over whether you are given a diameter or a radius. In the heat of an exam it is easy to make a slip, also be careful of units – make sure you are consistent.

10

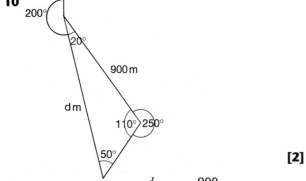

[2]

Using the sine rule $\frac{d}{\sin 110} = \frac{900}{\sin 50}$ [1]

So $d = \frac{900 \times \sin 110}{\sin 50}$ [1]

= 1100 m (to 2 s.f.) [1]

The use of the sine rule is strongly suggested here, as we effectively have a triangle in which we know two angles and one side. Once the sine rule has been set up, it just a matter of solving the resulting equation – and this is where most mistakes are made – it's best to 'cross-multiply': write the answers down and then divide to solve the equation, rather than try to do it in one step and get it wrong. However there is still partial credit for setting up the sine rule equation in the first place. Two significant figures is sensible, but 1104 might be allowed. With questions of this type, a sketch is

essential, particularly in seeing the different angles – it will show which angles you have found. If you have worked out an angle incorrectly but have shown it clearly, there are follow-though marks for correct working afterwards to be gained. A sketch also shows if the answer is 'reasonable' – even the rough sketch here suggests that something round about 1000 metres might be expected. Follow-through marks would be available in this question, for example, if you made a slip working out the triangle angles from the bearings.

Unit 4

1 (a) (i)

Working life, L, thousands of hours	Number of lamps	Frequency density
$1.6 \leq L < 1.8$	2	$2 \div 0.2 = 10$
$1.8 \leq L < 2.0$	4	$4 \div 0.2 = 20$
$2.0 \leq L < 2.1$	8	$8 \div 0.1 = 80$
$2.1 \leq L < 2.2$	4	$4 \div 0.1 = 40$
$2.2 \leq L < 2.6$	4	$4 \div 0.4 = 10$

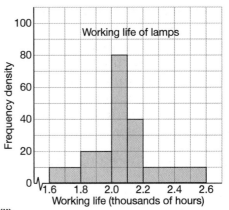

[3]

(ii) [2]

Working life, L, thousands of hours	Number of lamps (f)	Mid-value (x)	$f \times x$
$1.6 \leq L < 1.8$	2	1.7	3.4
$1.8 \leq L < 2.0$	4	1.9	7.6
$2.0 \leq L < 2.1$	8	2.05	16.4
$2.1 \leq L < 2.2$	4	2.15	8.6
$2.2 \leq L < 2.6$	4	2.4	9.6
Totals	**22**		**45.6**

Mean working life = 45.6 ÷ 22 = 2.1 thousand hours (to 2 s.f.) [2]

(b) (i) **[3]**

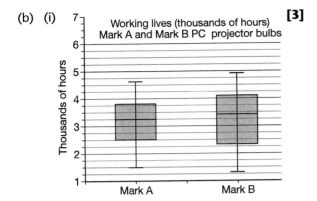

Working lives (thousands of hours)
Mark A and Mark B PC projector bulbs

(ii) Although the Mark Bs have a greater median working life, their figures are more spread out than the Mark As – so the Mark As are more reliable. **[2]**

The crucial relationship for drawing histograms is 'Frequency density = frequency in the particular class ÷ class width'. It is more efficient to add columns to the table and work out the frequency density of each class systematically as shown here – any slips, such as using the wrong class width, are soon apparent. You must know what the ends of the 'whiskers' (max and min values), the box ends (UQ and LQ) and the centre line (the median) represent. Comparing two box plots by giving one and asking you to draw the other and compare the two distributions is a popular question. Box plots can be vertical as here or horizontal. Probably their most important 'message' is about the spread in the distribution.

2 (a) P(arrive at correct time) = 1 – 0.05 – 0.1
$$= 0.85 \quad \textbf{[1]}$$
Number that arrive at correct time
400×0.85 **[1]**
= 340 trains **[1]**

(b) (i) P(two late) = P(late) × P(late)
$$= 0.1 \times 0.1 \quad \textbf{[1]}$$
$$= 0.01 \quad \textbf{[1]}$$

(ii) It has been assumed that the arrivals of the trains are independent events (if there were hold-ups due to something on the tracks this may not be the case). The multiplication rule can only be used for independent events. **[1]**

The relationship P(event happening) = 1 – P(event not happening) is a very useful tool in probability. Remember the 'sum rule' works only for mutually exclusive events and the 'product rule' works for independent events; trains being late may not be independent events, a break down or line hazard would have some effect on all the trains passing that area at about the same time. When asked to explain try to use the correct mathematical

language, most questions require a sentence or two, not an essay or a couple of words!

3 (i)

```
0|
0|8 9
1|0 0 0 0 0 0 1 1 1 1 2 2 3 3 3 4 4 4
1|5 5 5 5 5 5 5 5 6 6 6 7 7 7 7 8 8
2|0 2
2|5 7
3|
3|5 8
4|0
4|
```
$$5 \mid 2 = 52 \text{ m} \quad \textbf{[3]}$$

(ii) Most students were fairly accurate, but the trend was to over-estimate overall, with some students wildly over-estimating. **[1]**

When drawing a stem and leaf diagram you must give the key. The one here was about as difficult as can be asked, because making the stems go up in tens would not have shown much detail, so it was best to use two stems for each interval of ten, for example, 20, 21, 22, 23, 24, and 25, 26, 27, 28, 29. Remember the stem is the first part of the number. The digits in the leaves must be in numerical order.

4 (a)

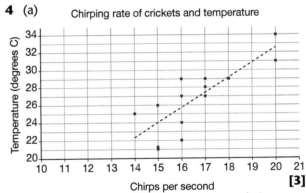

Chirping rate of crickets and temperature

[3]

(b) There is a fairly strong positive correlation between the two. **[1]**

Remember to title the chart and label the axes. Although it wasn't asked for, a trend line can help you see what's happening. Always try to judge how strong the correlation is, as well as saying if it is positive or negative. A common error, especially when the numbers to be plotted are similar, is to plot them the wrong way round; re-checking a couple of points against the table will reduce the chances of this.

5 (a) P(both white) = $\frac{11}{19} \times \frac{10}{18} = \frac{110}{342} = \frac{55}{171}$ **[2]**

(b) P(both white) + P(both black) **[1]**
$$= \frac{55}{171} + \left(\frac{8}{19} \times \frac{7}{18}\right) = \frac{55}{171} + \frac{28}{171} = \frac{83}{171} \quad \textbf{[1]}$$

(c) P(one black + one white) = P(black then white)
+ P(white then black) **[1]**

$$= \frac{8}{19} \times \frac{11}{18} + \frac{11}{19} \times \frac{8}{18} = \frac{88}{171}$$ **[1]**

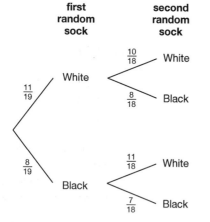

Although a tree diagram was not given, it certainly helped, if only to make clear that there was more than one way to get a black and a white; also with a tree diagram you can see that the branches sum to 1 – to prevent slips with the probabilities – especially when there is no replacement, as in this question. A slightly neater way to answer (c) is to use P(one black + one white) = 1 – P(both black) – P(both white).

6 (a) £20 000 to £29 999 **[1]**

(b) There are a total of 30.4 million so the median will be what the 15.2 millionth person has. This will be in the £15 000 to £19 999 class. **[1]**
Up to the start of this class there are (0.1 + 2.9 + 3.5 + 6.1) = 12.6 million
We need another 15.2 – 12.6 = 2.6 million. There are 5.1 million people in the next income class, so the fraction of this class needed is $\frac{2.6}{5.1} = 0.51$ (to 2 s.f.) **[1]**
This class interval is £5000 wide, so the median will be 0.51 into the class, which will be £(15 000 + 0.51 × 5000), which gives £17 550 as the median taxable income. **[1]**
Finding the median value for grouped data seems harder than it is – it just involves ratio to find where in the median class the median is located. Be careful to give a modal class or median class if asked. In this question the examiner said 'median' and so wanted the full calculation. This could have almost been guessed given the mark allocation of 3 – median class would probably only have been worth 1 mark.

7 (a)

Year	Quarter	Visitors	4-point moving average
2003	1	7	
	2	10	
	3	21	11.75 (= [7 + 10 + 21 + 9] ÷ 4)
	4	9	11 (= [10 + 21 +9 + 4] ÷ 4)
2004	1	4	10 (= [21 +9 + 4 + 6] ÷ 4)
	2	6	9.5 (= [9 + 4 + 6 + 19] ÷ 4)
	3	19	8.5 (= [4 + 6 + 19 + 5] ÷ 4)
	4	5	8.5 (= [6 + 19 + 5 + 4] ÷ 4)
2005	1	4	7.75 (= [19 + 5 + 4 + 3] ÷ 4)
	2	3	7.25 (= [5 + 4 + 3 + 17] ÷ 4)
	3	17	7.25 (= [4 + 3 + 17 + 5] ÷ 4)
	4	5	

[3]

(b) Number of visitors each quarter 2003 to 2005

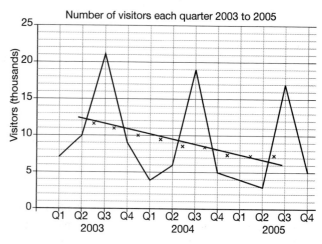

Number of visitors each quarter 2003 to 2005

[2]

(c) Trend line drawn on chart. **[1]**
The trend line shows a steady decline in numbers of visitors. **[1]**

Remember you can quickly calculate the next point in a moving average calculation by replacing the 'oldest' point by the next one (see above in the table). Entering points in a table like the above is useful as any rogue points (or errors) clearly show themselves.
Don't forget to plot moving averages between the quarterly values given – so the first point plotted was between Q2 and Q3 and 11.75 thousand visitors.
Your trend line should pass as near as possible to as many points as possible.
A common question is to ask you to complete a partially filled in graph or chart.

Index

Index